T0282317

Ecological Changes in the Zambezi River Basin

This book is a product of the CODESRIA Comparative Research Network.

Ecological Changes in the Zambezi River Basin

Edited by

Mzime Ndebele-Murisa
Ismael Aaron Kimirei
Chipo Plaxedes Mubaya
Taurai Bere

CODESRIA

Council for the Development of Social Science Research in Africa
DAKAR

ISBN: 978-2-86978-713-1

Typesetting: CODESRIA
Graphics and Cover Design: Masumbuko Semba

Distributed in Africa by CODESRIA
Distributed elsewhere by African Books Collective, Oxford, UK
Website: www.africanbookscollective.com

The Council for the Development of Social Science Research in Africa (CODESRIA) is an independent organisation whose principal objectives are to facilitate research, promote research-based publishing and create multiple forums for critical thinking and exchange of views among African researchers. All these are aimed at reducing the fragmentation of research in the continent through the creation of thematic research networks that cut across linguistic and regional boundaries.

CODESRIA publishes Africa Development, the longest standing Africa based social science journal; *Afrika Zamani*, a journal of history; the *African Sociological Review*; *Africa Review of Books* and the *Journal of Higher Education in Africa*. The Council also co-publishes *Identity, Culture and Politics: An Afro-Asian Dialogue*; and the *Afro-Arab Selections for Social Sciences*. The results of its research and other activities are also disseminated through its Working Paper Series, Book Series, Policy Briefs and the CODESRIA Bulletin. All CODESRIA publications are accessible online at www.codesria.org.

CODESRIA would like to express its gratitude to the Swedish International Development Cooperation Agency (SIDA), the Carnegie Corporation of New York (CCNY), the Andrew W. Mellon Foundation, the Open Society Foundations (OSFs), Oumou Dilly Foundation, Ford Foundation and the Government of Senegal for supporting its research, training and publication programmes.

Contents

Foreword

This is a fundamental book about a very important river basin located across eight countries in parts of the southern and east African region: The Zambezi River Basin. The approach adopted and developed for the Zambezi River Basin Catchment, using five river sub-catchments as case studies is innovative and with a systemic vision that takes into account the ecological conditions and the social dynamics. This approach was very much stressed in the Millennium Ecosystem Assessment; that is, by integrating the basic environmental functioning with uses of natural resources and local indigenous knowledge, it is possible to promote a full comprehension of the ecosystem services, their impact on the local and regional society and the impact of the multiple uses on the river ecosystem. Cultural differences, differing socio-economic systems, water and watershed governance are the key factors for future decisions and developments for strategic plans. The book thus has this dimension of an ecosystem dynamics description and a social, multiple use and economic analysis promoted by the various differing hydro-social characteristics of the five sub-catchments.

The methodology and review of the Comparative Research Method, with the discussion on the approach, limitations, use, is excellent. The conceptual framework of the study, theory and hypothesis are well connected and an adequate framework for such a large-scale, scientific, technical and management undertaking. Equally important are the descriptions on the sub-basins' functioning and conditions.

The description of the ecological changes, socio-economic status in four countries and the major drivers and pressures in the Zambezi Basin is well articulated, precise and adequately informative. The discussion on the sustainability model of Venn is very welcome and relevant in this context. The information on climate changes, flow alternations, water pollution and poverty are also important, well-articulated and place the roots for the subsequent chapters.

A fundamental contribution of the book is in the river health assessment in east and southern Africa. Comparative uses of indicators, sampling of biota analyses, are of prime interest and are well described and designed, and methodologically coherent. The future climate scenarios of the Zambezi River Basin, as placed in the book help to understand (again with a systemic approach) the integration of the impact of climate change on the ecosystem and on the society. These processes are interdependent. The adaptive capacity of the communities in the river basin catchment of the Zambezi is a highlight of this book and has a very high strategic value.

Equally important is the description of the hydrological dynamics of the Zambezi River Basin and the trend analysis of the discharge. Ecosystems services of the rivers depend upon the flow intensity, variations and tendency. This is the fundamental asset for considering then, the multiple uses, the intensity of the hydrological cycle and the impact of climate change. The environmental flow analyses are within this framework and conceptual basis. River flows, social processes, impacts and economic development are all integrated locally and regionally.

The whole book has a strategic concept considering integrated water resource management. The description and analysis, the synthesis, the integration of concepts are relevant approaches to design new developments and to promote economic progress with high-level decision-making and prioritization on the environmental issues. The book has a great number of references and can be extremely useful worldwide due to the regional importance of the Zambezi River; but also by its innovative and creative approaches that advance environmental sciences at all levels.

Prof Dr José Galizia Tundisi
International Institute of Ecology and
Environmental Management
São Carlos, SP, Brazil
30 May 2017

Acknowledgments

This book project is an output of the CODESRIA Comparative Research Network Grant (2013-2015)-funded project (CRN/CTR3/2013). We acknowledge funding from CODESRIA which covered the project on which the book is based. CODESRIA also provided blind reviewing, copy and technical editing, which we are grateful for as they greatly enhanced the quality of the book. The Chinhoyi University of Technology (CUT) is duly acknowledged for co-funding writing workshops. A contribution by Ms Olga Kupika, Mr Beaven Utete, Mr C. Mapingure and Mr C. Muzari (Chinhoyi University of Technology) of a case study of the Mupfure River in Chapter 5 is duly acknowledged. The assistance of Olga Kupika, Regean Mudziwapasi, Rutendo Musimwa and Varaidzo Chinokwetu with field surveys in Rafingora in June 2014 is acknowledged.

The editors also acknowledge the assistance of the following persons during the book project: Prof Linda Hantrais (Centre for International Studies, London School of Economics), Dr Patience Mutopo (Centre for Development Studies, Chinhoyi University of Technology), Dr Mathias Igulu (Tanzania Fisheries Research Institute), Dr Crispen Phiri (Department of Freshwater and Fishery Sciences, Chinhoyi University of Technology), Dr Charles Musil (retired Senior Scientist-South African National Biodiversity Institute), Dr Victor Kongo (Department of Water Resources and Engineering, University of Dar es Salaam) and Dr Loreen Katiyo (World Wide Fund for Nature-WWF, Zambia), all of whom were Specialist Chapter Reviewers.

Professor José Galizia Tundisi (University of São Carlos, Brazil) was gracious enough to read the book draft and to provide the insightful foreword. We are also indebted to Mr Semba Masumbuko for several professional services that include typesetting, provision of and quality assurance of all the diagrams in the book as well as technical editing.

This book is a result of multi-faceted collaboration efforts, many sleepless nights and assiduous workings across the Zambezi region. We are grateful for the funding and other assistance (weare not able to mention everyone) that were rendered during the undertaking of the project, book writing and printing. Without this effort and goodwill, this book would not have been possible.

List of Boxes, Figures, Plates and Tables

List of Boxes

List of Figures

List of Plates

List of Tables

List of Abbreviations

AEHRA	Adapted Ecological Hydraulic Radius Approach
ANOVA	Analysis of Variance
ASPT	Average Score Per Taxon
BFPS	Barotse Flood Plain System
BFI	Base Flow Index
BWB	Basin Water Boards
BBM	Building Block Methodology
CAP	Canonical analysis of principal coordinates
CCA	Canonical correspondence analysis
CC	Catchment Councils
CMCs	Catchment Management Committees
COD	Chemical Oxygen Demand
CV	Coefficient of Variation
CRM	Comparative Research Method
CORDEX	Coordinated Regional Climate Downscaling Experiment
DRV	Desktop Reserve Model
DDT	Dichloro-Diphenyl-Trichloroethane
DO	Dissolved Oxygen
DRIFT	Downstream Response to Imposed Flow Transformation
DPSIR	Driving Forces-Pressures-State-Impacts-Responses
ELOHA	Ecological Limits of Hydrological Alteration
ENSO	El Niño Southern Oscillation
EC	Electrical Conductivity
ESCOM	Electricity Supply Corporation of Malawi
EFA	Environmental Flow Assessment
EFMs	Environmental Flow Methodologies
EFRs	Environmental Flow requirements
EF	Environmental Flows
EWR	Environmental Water Requirement
FAII	Fish Assemblage Index
FDC	Flow Duration Curve
GFDL-CM	Geophysical Fluid Dynamics Laboratory Coupled Model

GCMs	Global Circulation Models
GDP	Gross Domestic Product
HFSR	Habitat Flow Stressor Response
HQI	Habitat Quality Index
HRWL	Highest Regulated Water Level
HAT	Hydrological Assessment Tool
HI	Hydrological Index
IVA	Impact and Vulnerability Assessments
IHA	Indicator of Hydrologic Alteration
IKS	Indigenous Knowledge Systems
IFIM	Instream Flow Incremental Methodology
ILM	Integrated Landscape Management
IWRM	Integrated Water Resources Management
IWRMD	Integrated Water Resources Management and Development
IPCC	Intergovernmental Panel on Climate Change
IUCN	International Union for Conservation of Nature and Natural Resources
IPA	Interpretive Phenomenological Analysis
ITCZ	Inter-Tropical Convergence Zone
LA	Landscape Approach
LIKSP	Local Indigenous Knowledge and Practices
MAR	Mean Annual Runoff
MSM	Meteorological Services of Malawi
MSZ	Meteorological Services of Zimbabwe
MCM	Million Cubic Metres
MoIWD	Ministry of Irrigation and Water Development
NASS	Namibian Scoring System
NAWAPO	National Water Policy (Tanzania)
NWRA	National Water Resources Authority
NRM	Natural Resources Management
NGO	Non-Governmental Organization
OKAS	Okavango Assessment System
PRA	Participatory Rural Appraisal
PoF	Percentage of Flow
PHABSIM	Physical Habitat Simulation
R:E	Rainfall-Evaporation Ratio
RVA	Range of Variability Approach
RCMs	Regional Circulation Models
RHA	River Health Assessment
SRP	Soluble Reactive Phosphate
SADI	South African Diatom Index

SASS	South African Scoring System
SADC	Southern African Development Community
SADCC	Southern African Development Co-ordination Conference
SCC	Sub-catchment Councils
SSA	Sub-Saharan Africa
SBA	Sustainability Boundary Approach
TMA	Tanzania Meteorological Authority
TARISS	Tanzania River Scoring System
TDS	Total Dissolved Solids
TH	Total Hardness
TN	Total Nitrogen
TP	Total Phosphate
TDI	Trophic Diatom Index
WRA	Water Resource Areas
WRIS	Water Resources Investment Strategy
WRM	Water Resources Management
WRMA	Water Resources Management Authority
ZAB	Zaire Air Boundary
ZAMCOM	Zambezi Commission
ZRA	Zambezi River Authority
ZRB	Zambezi River Basin
ZISS	Zambia Invertebrate Scoring System
ZMTR	Zambia Macrophyte Trophic Ranking System
ZMD	Zambia Meteorological Department
ZINWA	Zimbabwe National Water Authority

Editors and Contributors

Editors

Mzime Ndebele-Murisa is a Freshwater Ecologist who was based in the Department of Freshwater and Fishery Sciences, School of Wildlife, Ecology and Conservation at Chinhoyi University of Technology during the writing of this book. Currently, she is a Program Specialist at START International in charge of the Future Resilience for African Cities and Lands (FRACTAL) and other programs. Mzime's research interests include aquatic ecology, fishery sciences, plankton ecology, climate modelling, adaptation, vulnerability, and impact assessments, as well as water resources management, sustainability sciences and development. Mzime has been involved in several trans-disciplinary research and development programmes including the IDRC-funded African Climate Change Fellowship Programme (ACCFP), the DFID-funded Climate Implementation Research Capacity Leadership Enhancement (CIRCLE), the NERC-funded as well as the World Climate Research Programme-commissioned Coordinated Regional Climate Downscaling Experiment (CORDEX Africa) in the role of a Trainer and Evaluator, Principal Investigator and Mentor, Coordinator, and Research Team Member. Mzime has a BSc (Honours) in Biological Sciences and an MSc in Tropical Resource Ecology from the University of Zimbabwe and attained a PhD degree from the University of the Western Cape, South Africa.

Ismael Aaron Kimirei is an Aquatic ecologist with extensive experience in fish ecology, limnology, aquatic resources management, and climate change issues. Ismael has more than 16 years of working in the great lakes of East Africa and has authored and co-authored 35 peer-reviewed journal articles. His Master's research focused on how limnological parameters affect fish catches in Lake Tanganyika, while his Doctorate research studied the nursery function of coastal habitats to coral reef fishes in Tanzania. Ismael has worked with a consortium of Belgian universities and institutes to study how Lake Tanganyika and its fisheries respond to climate change. His current research focuses on how climate change in Lake Tanganyika is affecting the fisheries of the lake and community livelihoods. He is currently the Director-General of the Tanzania Fisheries Research Institute.

Previously, he worked as the Centre Director for the same organization in Kigoma for the past decade. Ismael holds BSc (Hons) and MSc degrees from the University of Dar es Salaam and a PhD in Ecology from the Radboud University Nijmegen, The Netherlands.

Chipo Plaxedes Mubaya holds a PhD in Development Studies and is currently the International Collaborations Manager at Chinhoyi University of Technology, Zimbabwe. She has conducted livelihoods and vulnerability assessments, gender analysis in CBNRM and climate change adaptation and institutional and stakeholder analysis in Sub-Saharan Africa. Chipo has worked with the International Development Research Centre (IDRC) and the DFID-funded African Climate Change Fellowship Programme both as a Doctoral Fellow and as the Senior Programme Officer at the Institute of Resource Assessment, University of Dar es Salaam. Chipo was awarded a World Social Science Fellowship in Sustainable Urbanization by the International Social Science Council (ISSC) in 2013 and has been involved in applied research, specifically on enabling rural innovations through linking farmers to markets and understanding social dynamics for farmers in participatory action research. Chipo has also been involved in engaging local-level stakeholders in a multidisciplinary setting in various development-oriented research projects both as Principal Investigator and co-researcher. Her research interests are in rural development, agriculture, and natural resource management and rural/urban climate change adaptation and decision making.

Taurai Bere is a Freshwater Ecologist with a passion for excellence which is driven by the desire to continue to grow in knowledge and leadership, excel in applied aquatic ecology, interact and share with regional and international colleagues, and develop excellent solutions to challenges facing freshwater resources. His main areas of expertise (with extensive publications) include river health assessment, biological monitoring of aquatic systems, aquatic ecotoxicology, and integrated water resources management. He is also interested and has knowledge in climate change impacts, ecological modelling of freshwater systems and bioremediation of aquatic systems. Currently, he is advising a group of researchers working on a locally funded research project entitled, 'Building adaptive capacity to mitigate effects on climate change on ecosystems and livelihoods dependent on rivers in Mashonaland West Province: case study of Makonde and Sanyati District'. He graduated with a PhD degree in Ecology and Natural Resources from the Federal University of São Carlos-SP, Brazil in June 2011; has an MSc. degree in Tropical Hydrobiology and Fisheries (2003-2005) and a BSc Honours degree in Biological Sciences from the University of Zimbabwe.

Contributors

Geoffrey Chavula is an Associate Professor in Water Engineering at the University of Malawi-The Polytechnic and is also a former Director of the Centre for Water, Sanitation, Health and Technology Development (WASHTED). He teaches undergraduate courses in Engineering Hydrology, Hydraulics, and Water Resources Management in the Department of Civil Engineering. He has a PhD in Water Resources Science from the University of Minnesota, USA; a Master of Science degree in Water Engineering from the University of Newcastle, England; and a Bachelor of Science degree in Physics and Earth Sciences from the University of Malawi.

Lulu Tunu Kaaya is a Lecturer in Freshwater Ecology in the Department of Aquatic Sciences and Fisheries Technology at the University of Dar es Salaam, Tanzania where she teaches limnology and watershed management. She is also a lecturer in stream ecosystem concepts and stream integrity assessment in the International joint Master's degree in Limnology and Wetland Management Programme (Boku University-Austria, Egerton University-Kenya and UNESCO-IHE, The Netherlands). Dr. Kaaya has been doing research in the rivers and lakes of Tanzania for 15 years. Her research focus is around the ecological role and functions of freshwater ecosystems in the management of water resources. She has been providing professional consultation on the integration of ecology into watershed management approaches in Tanzania, i.e., Environmental Flow Assessment, River Health Assessments and establishment of integrated water resources management and development plans. Her Ph.D. thesis titled "Biological Assessment of Tropical Riverine Systems Using Aquatic Macroinvertebrates in Tanzania, East Africa" provides a detailed analysis of bioassessment from the regional perspective. She has published nine peer-reviewed articles and conducted several outreach programmes on bioassessment.

George V. Lugomela is a seasoned Hydrologist working with the Ministry of Water and Irrigation in Tanzania. He has worked with the Ministry since 2000 where he later served as an Assistant Director responsible for Water Resources Planning, Research and Development in from 2012. In 2019, he was appointed the Director of Water Resources Department of the Ministry of Water and Irrigation. He has participated in several water resources management projects, most notably the Pangani River Basin Management Project (PRBMP) which focused on flow assessment and environmental flow analyses in the basin. He supervised the preparation of Integrated Water Resources Management and Development (IWRMD) plans for different basins in Tanzania which are the blueprints for holistic sustainable water resources management and development. Dr Lugomela has specialized in groundwater hydrology, particularly numerical modeling of groundwater flow and contaminant transport using the Finite Element Method where he has several publications. He also lectures on a part-time basis in Groundwater Hydrology and Modeling at the University of Dar es Salaam and the Nelson Mandela African Institute of Science and Technology (NMAIST).

Rashid Tamatamah is a Freshwater Ecologist and has served as a Senior Lecturer and Head of the Department of Aquatic Sciences and Fisheries Technology at the University of Dar es Salaam, and the Director General of TAFIRI. His research interests include nutrient dynamics in aquatic systems, fish biodiversity and fisheries management, and construction and management of aquaculture systems. He has previously undertaken the first field based study for estimating atmospheric phosphorus deposition in Lake Victoria and has been a Lead Scientist in a number of studies on fish biodiversity assessment and fisheries for Environmental Flow Assessments (EFAs), EIAs and Audits of mining and hydropower projects in Tanzania and Kenya. He has also participated in the preparation of the Integrated Water Resources Management and Development Plans (IWRMDP) for several River Basins of Tanzania and conducted a review of the General Management Plan for the Mnazi Bay-Ruvuma Estuary Marine Park (MBREMP). Dr. Rashid Tamatamah is currently the Permanent Secretary at the Ministry of Livestock and Fisheries, Tanzania.

Tendayi Maravanyika holds a PhD in natural resources management from Wageningen University in the Netherlands. She obtained an MSc degree in Management of Agriculture Knowledge and Information Systems from the same university. Tendayi also has a BSc Agriculture Honours degree (with specialisation in Agricultural Economics) from the University of Zimbabwe. For more than fifteen years, Tendayi has worked on designing and implementing projects for communities to adapt to change (including climate change) in the following fields: agriculture, forestry and fisheries. She has worked in several African countries including Ghana, Zimbabwe, Zambia, Uganda and Ethiopia, and has assisted several teams to design and implement projects, enhance community resilience and adaptive capacity. Tendayi has strong interest in the following topics: natural resources management, participatory action research, adaptive collaborative management, climate change adaptation and community-based natural resources management.

Masumbuko Semba is a scientist who works at the Nelson Mandela African Institution of Science and Technology and is currently a Doctoral student at the University of Dar es Salaam in Tanzania. Semba is specialized in earth observation using geographical information systems and remote sensing technology to tackle science-based questions facing society. Semba is currently working with NOAA/

AOML's Physical Oceanography Division (PhOD) through the global drifters programme. In this project, Semba is testing several aspects of global ocean observing systems by combining satellite altimetry and drifter observations to uncover the ocean circulation dynamics in the western Indian Ocean region.

Tongayi Mwedzi is a Conservation Ecologist based in the Department of Wildlife, Ecology and Conservation at Chinhoyi University of Technology. Tongayi has special interests in biomonitoring and hydro-ecology in which he has authored and co-authored a number of papers in international, peer-reviewed journals. He is also knowledgeable regarding climate change impacts and modelling. His current research focuses on modelling environmental flow requirements and biotic response on streams impacted by anthropogenic activities and climate change. He holds a BSc (Hons) degree in Forest Resources and Wildlife Management from the National University of Science and a DPhil from Chinhoyi University of Technology, Zimbabwe.

Tinotenda Mangadze was previously a Teaching Assistant in the Department of Freshwater and Fishery Sciences at Chinhoyi University of Technology (2013- 2016). Her research interests lie in the area of River Health Assessments, Biological Monitoring and Ecotoxicology. In recent years, she has concentrated on research on the response of biotic assemblages to changes in water quality in different land-use settings. Tinotenda is a fellow of the Organization for Women in Science for the Developing World (OWSD) and she is also involved in organizations such as the Southern African Wildlife Management Association (SAWMA) and is currently based at University of Rhodes, South Africa.

Judith Natsai Theodora Kushata has been involved in several research projects, including but not restricted to, both natural and social sciences. Judith is interested in a wide range of issues in applied ecology and conservation, principally in the natural science field. She has found her research passion in the context of rural development and socio-ecosystems, and received a grant from the Rufford's Trust to carry out a study on human-wildlife interactions, with a focus on the pangolin, which is biologically interesting and poses important questions for conservation in the domains of trade and sociology as well as ecology. She realized the importance of using the holistic management approach; a trait which got her involved in various social projects including the OXFAM Climate Change Adaptation (CCA) and Disaster Risk Reduction (DRR) project in Zimbabwe. Judith completed an MSc at Stellenbosch University in South Africa, studying the phylogeography and diversity of Anurans in the Wild Coast forests (Eastern Cape) and macro and micro-evolutionary phenomena. She is currently working as a Program Coordinator at SIVIO Institute, a policy think tank in Harare, Zimbabwe.

Sandra Zenda is an emerging Conservation Ecologist who kick-started her career at Chinhoyi University of Technology and Kyle Recreational Park in Zimbabwe. Her major research interests lie in aquatic ecology, especially biomonitoring of river systems and how climate affects river systems and livelihoods. In an effort to bridge the gap between scientific reporting to peers and communication to the lay public, she has been actively involved in science communication. Sandra has worked as a Research Assistant under a World Bank-funded project in Harare and is currently completing an MPhil at Chinhoyi University of Technology under the CDKN-Df ID-NERC funded Future Climate for Africa's Future Resilience for African Cities and Lands (FRACTAL) Project that focuses on the nexus between climate, water and energy for the city of Harare in Zimbabwe. She is also a Program Coordinator at Institute for Young Women Development, an NGO based in Bindura, Zimbabwe.

1

Introduction and Background

Taurai Bere, Mzime Ndebele-Murisa, Chipo Plaxedes Mubaya and Ismael Aaron Kimirei

Introduction

The river systems of east and southern Africa's semi-arid areas are under severe stress due to several factors among which are climate change and variability as well as explosive population and urban growth. In these regions, water demand is outstripping supply, with water withdrawals being estimated to have increased six-folds during the twentieth century (IHE Delft 2002; Vörosmarty et al. 2005, Molden 2007). Widespread degradation of freshwater ecosystems has been reported in the region, with water quality declining due to increased pollution among other factors (Nel et al. 2011). This degradation of freshwater ecosystems inevitably compromises ecosystems service delivery and results in expensive management interventions and the ecosystems' loss of resilience to the changing circumstances (Nel et al. 2011). In addition, increases in frequency and intensity of extreme weather events such as droughts and floods are complicating water resources management in the African continent and the southern African region in particular (Chigwada 2005; IPCC 2007, 2014, Darwall et al. 2009; Dube and Chimbari 2009; Mwendera 2010).

Extreme weather events bear potential negative implications on river flows, possibly compro-mising the ecological integrity of rivers and hence the resultant ecosystem goods and services relied upon by surrounding communities. The large extent to which societies rely on aquatic ecosystem goods and services has provided the impetus for assessments of ecological changes and adaptive capacity in the livelihoods of mostly rural communities. This book is the first of its kind to provide a synthesis of ecological conditions and their downward impacts on societies in a comparative research network across five river catchments within the Zambezi Basin and which lie in four basin countries namely; Malawi, Tanzania, Zambia and Zimbabwe. Thus, the book can be seen as contributing to a research gap and providing research-based evidence for the formulation of management policies and strategies that are useful in the decision-making process of a range of stakeholders, including government departments and conservation organizations.

The Zambezi River and its dense network of tributaries discharge an average of 2,600 cubic metres per second (m^3s^{-1}) of water, a rate in the same range as the Nile (2,830 m^3s^{-1}) and the Rhine (2,200 m^3s^{-1}) Rivers (Beck and Bernauer 2010). The basin has abundant water, fertile land and soils for agriculture, mineral rich deposits and diverse habitats that are home to large populations of wild-life (SARDC et al. 2012). This natural capital defines the basin's economic activities which include agriculture, forestry, manufacturing, mining, conservation and tourism. The basin is also a centre for scientific monitoring and research. The most prominent wetlands in the basin are Barotse Flood Plains and Bangweulu in Zambia; Okavango Delta that is shared between Botswana and Namibia; the Chobe Swamps in north-eastern Namibia; the Linyanti Swamp in Botswana; the Busanga Swamps on the Lunga River, the Lukanga Swamps and the Kafue Flats on the Luangwa River in Zambia; and the Elephant Marsh near the town of Chiromo in Malawi, as well as Kazuma Pan in Zimbabwe (SARDC et al. 2012). The Zambezi River drains part of eight countries and is the most shared river system in southern Africa. The Zambezi riparian countries include Angola, Botswana, Malawi, Mozambique, Namibia, Tanzania, Zambia and Zimbabwe. The Zambezi River Basin has witnessed drastic changes in its natural environment in recent years mainly as a result of demo- graphic dynamics, urbanization and increasing demand for agricultural land among other factors. These drivers have brought about changes to ecosystems, water resources and the way different cultures interact.

Overview of Ecological Changes and Socio-economic Status

Although the book is about the Zambezi River Basin which spans eight countries, we concentrate on five river basins contained within four of the Zambezi riparian countries. These four countries' natural resources and demographies are described briefly in the ensuing sections.

Malawi

Malawi is located in south-east Africa and is bordered by Tanzania to the north, Zambia to the west, and Mozambique to the south. The country's current (2016) population is an estimated 17 million that inhabits the country's total area of 118,000 square kilometres, 20 per cent of which is covered by water (rivers and lakes) (BBC News 2015). The country has a sub-tropical climate, ranging from semi-arid in the Lower Shire Valley, semi-arid to sub-humid on the plateaus and sub-humid in the Highlands. Of the total land area, 31 per cent is suitable for rain-fed agriculture, 32 per cent is marginal and 37 per cent is unsuitable for agriculture. Malawi is among the least developed countries in the world, with an estimated Gross Domestic Product (GDP) of $12.81 billion, and a per capita GDP of $900 in 2009, and inflation estimated at around 8.5 per cent (The World Factbook 2010). Malawi's economy is driven by agriculture, and smallholders contribute more than 80 per cent of Malawi's agricultural production, which is dominated by maize, the country's staple food crop (The World Factbook 2010). The well-being of the majority of Malawians is highly vulnerable to climate change because of the country's dependence on rain-fed agriculture.

Tanzania

Tanzania has over 45 million people and about 44 million ha of arable land; and is located in east Africa. Because of its diverse ecosystems, topography and climate, the country is endowed with rich natural resources and biological diversity. Apart from the vast areas of arable land, she has extensive forest and wildlife resources, rangelands, aquatic resources and minerals (URT 2001). Over 60 per cent of the total export earnings come from agriculture, livestock, forestry and fisheries. These sectors also contribute to over 65 per cent of the country's GDP and employ over 80 per cent of her population (URT 2014). Tanzania has an estimated forest cover of about 55 per cent of the total land area, which is currently being degraded at an alarming rate. Despite this fact, over 90 per cent of energy consumption in the country is contributed by forests. Hydro-power, on the other hand, provides about 37 per cent of the total energy share in the country. For a country with such a reliance on natural resources for her economy, ecological changes would bear adverse consequences. Ecological changes cause loss of habitats and biodiversity, loss of freshwater integrity, threats from pollution and over-exploitation of resources. It has already been shown, for example, that fisheries production is in decline as a result of both natural and anthropogenic stressors (O'Reilly et al. 2003; Witte et al. 2009; see also Chapter 3 of this book).

Zambia

Zambia has an estimated population of 16.5 million and extends over an area of 752,618 km². About 80 per cent of Zambia is covered by miombo vegetation type whilst the other vegetation types are in the form of grasslands and forest areas. The country's woodland's coverage is predominantly determined by precipitation and altitude [Ministry of Tourism, Environment and Natural Resources (MTENR 2011)]. The Zambezi River catchment, which occupies three quarters of the country, has three major sub-basins (i.e., Kabompo, Kafue and Luangwa) stationed within the country. Zambia, which recognizes the crucial role tourism plays in economic development, has a looming tourism industry which hinges primarily on two assets, namely the Mosi-oa-Tunya (Victoria) Falls and its rich array of wildlife protected areas (Liu and Mwanza 2014). The country has a population annual growth rate of 2.9 per cent and centres on five main industries (manufacturing, agriculture, construction, transport and mining) which are the main sectors of its economy (MTENR 2011). As reported by the GoZa (2012), the poverty status of rural communities in Zambia is very high, hence catapulting high dependency on natural resources which fall under the agriculture industry at the expense of their persistence. Zambia's MTENR (2007) recognized that climate change could have drastic impacts on the livelihoods of communities who largely depend on natural resource usage if left unaddressed.

Zimbabwe

Zimbabwe is a landlocked country which is in the southern African region. The country is bordered by South Africa, Zambia, Mozambique and Botswana (Brown et al. 2012a) and covers an area of 390,757 km². Forty per cent of the country is covered by woodlands and forests (FAO/WFP 2008), with 95 per cent of the forest cover being Savannah woodland vegetation type (Nyamadzawo et al. 2013). The country's population was a record 13 million by 2011 and is currently estimated at 15,764,255, with approximately 62 per cent of the population living in rural areas (Brown et al. 2012). Zimbabwe is renowned for its high dependency on agriculture as well as other resources which are climate-sensitive and therefore rely on natural rainfall (Chagutah 2010). The country is faced by rapidly declining economic productivity which can be related to declining agricultural outputs, which has in turn led to increasing overexploitation of natural resources for survival (Brown et al. 2012). Zimbabwe's GDP was estimated at $819.42 in 2016 (WorldBank 2015). The country's agro-ecological regions have shifted dramatically as a result of the rapidly changing climate, thus negatively affecting the livelihoods of the native people and the country (Brown et al. 2012; Mugandani et al. 2012; Manyeruke et al. 2013; Ndebele-Murisa and Mubaya 2015).

Healthy River Ecosystems and the Concept of Sustainability

Rivers are not only flowing bodies of water but they also thrive with life and sustain ecosystems. The water contained therein directly or indirectly affects all facets of life, activity and aspirations of human society. River systems provide for many of our fundamental needs such as water for drinking and irrigation, industrial, aesthetic and recreational purposes, food such as fish, numerous life-sustaining ecosystem services such as water purification, carbon sequestration, prevention of floods and the easing of droughts. Some of the best evidence of the importance of rivers to humans lies in the fact that first known civilizations were mostly located near larger waterways. To date, rural households in Africa are mainly located along rivers, clearly demonstrating the importance of rivers as resources for the sustenance of local livelihoods.

However, rivers can only provide our fundamental needs if they are clean and healthy. Otherwise, they become useless as a source of water and food for humans, and for other ecological, social or cultural value. Thus, the importance of clean, healthy river ecosystems cannot be overemphasized. It is crucial to look after our river systems and protect their resilience and ability to recover from natural and man-made disturbances in order for them to maintain their ecosystem functions.

Wise utilization of our rivers is critical to the sustainable development of emerging economies and the well-being of all citizens, particularly the poorest, who depend directly on the health of natural resources for their livelihoods (Millennium Ecosystem Assessment 2003; Nel et al. 2011). With economic, social and ecological systems being intricately woven, healthy ecosystems, social and economic development is not possible (Figure 1.1).

Despite the use of a Venn diagram in depicting sustainable development in Figure 1.1, we, however, advocate for the Russian doll sustainability model (see Chapter 2, Figure 2.7) because of its more inclusive nature. Within the chapters, we utilize the Driving Forces-Pressures-State-Impacts-Responses (DPSIR) model to explore the themes and sub-themes therein. Chapter 3, which entails ecological changes of the Zambezi River basin uses the DPSIR model in great detail, with the rest of the chapters following suit. The drivers and pressures are discussed in the introductions of the chapters while the status and impacts come through within the analytical sections of the chapter. Also, where data were available, detailed analyses were employed to investigate the nature of the status and impacts of ecological changes. This is especially true of all the chapters, particularly with the use of desktop studies and secondary data; but more so for Chapters 3, 4, 5 and 6 where quantitative data were employed to provide empirical evidence of trends using statistical analyses. The responses within the DPSIR model are explored in the discussion sections of the chapters with recommendations for policy and institutional development (formulation, implementation and re-formulation) discussed as a way forward. Thus, a synthesis of ecological changes and their downward impact on societies in the context of sustainable development are presented and our hope is that such an approach then captures the issues raised and concurrently attempts to address them in a holistic manner.

The Zambezi River is no exception to the application of sustainability and, if applied, the ecosystem can function at sustainable capacities. The total number of people living within the Zambezi River basin is estimated to be around 32 million and a huge number depend directly on the basin's resources. It has been estimated that more than 60 per cent of riparian communities in and around the basin depend on fisheries for livelihood and on fish as the sole source of protein

(Ndebele-Murisa et al. 2010; Tweddle et al. 2015). The river basin is rich in natural resources. Main activities focus on fisheries, mining, agriculture, tourism and manufacturing. Beilfuss and dos Santos (2001) explain that the 'predictable' hydrological regime of rivers is the core frame that supports local livelihoods as the local people take advantage of the annual flooding and use the fertile flood plains to practice recession agriculture, grazing livestock in flood plains, fishing and harvesting of other natural products.

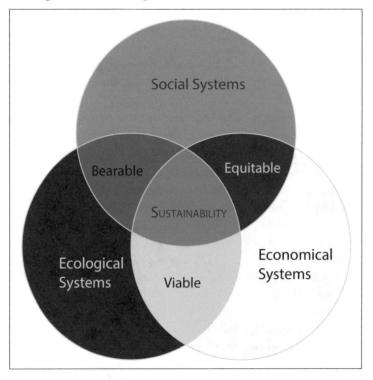

Figure 1.1: Venn diagram sustainability model (Modified from Bruntland) (1998)

Despite the Zambezi River Basin's importance, very few river assessments have been conducted in the basin. In addition, no comparative studies have been conducted for the basin to dertermine the factors affecting river flow and ecosystem functioning and therefore ecosystem goods and services, as well as to establish the extent to which the reliant communities' coping strategies to ecological changes are similar. This book seeks to redress some of the research gaps identified in this era. The lessons learnt from this study can be used to the advantage of communities in and around the Zambezi River.

Major Drivers and Pressures in the Zambezi Basin

The Zambezi River is the largest river in southern Africa and its dense network of tributaries and associated ecosystems constitute one of the region's most important natural resource systems (Chenje 2000). Despite this, sustainable utilization of the Zambezi River Basin suffers from a number of threats. Ecological changes in the Zambezi River systems are already underway. Many drivers and pressures contribute to the degradation of the Zambezi River ecosystems; compromising ecosystem goods and services provision, and it is difficult to isolate a single cause. Among the threats to the Zambezi River Basin ecosystems are climate change and variability, flow alterations, poverty and water pollution.

Poverty

The economies of the Zambezi riparian countries are generally characterized by low levels of industrial development and economic growth. The existing industries strongly depend on the ecosystem goods provided by the river and/or its tributaries. A considerable part of the people living in the river basin area are very poor, only little educated, and have limited or no access to education and health facilities, which is mainly a consequence of the historical backgrounds of these countries. In the Manyame and Sanyati areas (Zimbabwe) and along the Kafue River (Zambia), for instance, some of the people have been forced to turn to alluvial gold panning as the area is agriculturally unviable (Shoko 2002). Fishing forms the core of the people's livelihoods, as was found in the case study along the Manyame River, where land-based capital was the main source of livelihoods before (see Chapter 3 of this volume). Any disturbance, therefore, to not only the Zambezi but any river's flow and ecosystem bears direct impact on the communities' livelihood and welfare.

Water Pollution

Water resources in the Zambezi Basin are threatened by pollution, posing health risks to the societies depending on these resources. Pollution sources include industrial and mining effluent, agricultural pesticides and fertilizers, and domestic effluent including sewage. This is increasingly becoming serious, especially as failing water treatment infrastructure battles to treat the increasing domestic and industrial effluent from towns and cities. The end result is a compromised structure and function of the river ecosystems and hence the resultant ecological goods and services and associated livelihoods. These changes ardently affect river ecosystems. Keeping freshwater systems healthy will help them adapt to these changes with least disruption to ecosystem services. In this regard, it is therefore possible to systemically monitor our aquatic systems and their integrity and health in order to surmise their ecological functioning and ability to provide precious services to biological organisms and the human populace. Different organisms have been used as biological indicators of river ecosystem health in east and southern Africa. These assays having mostly emanated from South Africa have been further developed and contextualized for other specific river basins in the region (see Chapter 4 of this book).

Climate Change

The use of Regional Circulation Models (RCMs) to project future climate scenarios is gaining momentum. However, there is a need to improve understanding and modelling of climate changes related to the hydrological cycle at scales relevant to decision-making (Bates et al. 2008). This is because large-scale assessments often have a limited shelf life and should be replaced with new studies at regular intervals. At the regional scale of the Zambezi River Basin, there is evidence of a broadly coherent pattern of change in annual runoff, with some regions experiencing an increase in runoff (e.g., high latitudes) and others experiencing a decrease (Bates et al. 2008; Rummukainen 2010; Samuelsson et al. 2011). In the semi-arid and arid areas of the Zambezi River Basin, rainfall is projected to decrease with a greater frequency of droughts expected, causing the areas to become drier and hotter (Unganai 1996; Arnell 1999; Vörösmarty et al. 2000; Arnell 2004; Vörösmarty et al. 2005; Kusangaya et al. 2014). Records from the countries that are in the Zambezi Basin reflect that temperatures have risen by over 0.5°C in the past 100 years, with the last decade recording the warmest and driest period ever. Over the past 20 years, there has been noticeably less rainfall and drought has become an increasingly serious threat (Tumbare and Zambezi River Authority 2004; IPCC 2007; Shongwe et al. 2009, 2011; IPCC 2014).

Flow Alterations

The resulting response of ecological processes to hydrological processes is a critical issue in the assessment of environmental flow requirements (Sun et al. 2012). With water demand outstripping supply, water withdrawals are estimated to have increased six-fold during the twentieth century (Molden 2007). Climate-induced flow alterations, coupled with some water diversion and abstraction, are common in the Zambezi River Basin. Increased temperatures, coupled with increased rates of evapo-transpiration, are likely to have disproportionately large impacts on runoff and river flow (Schulze 2005), with implications for the planning and management of water resources. Already, the past decade has seen the occurrence of some serious droughts and their increased frequency around the basin with the corroborated reduced rainfall, resulting in reduced water flows and therefore water supply for services such as environmental flows and hydroelectricity for instance. The major hydropower stations in the basin (Kariba, Cabora Bassa and Shire) attest to this as their power generation oscillates with the Zambezi and Shire River water flows respectively. The anthropogenic (e.g., abstraction, building of dams) or natural (mainly climate-induced) alterations in flow regime (e.g., timing, frequency, speed or volume of flow) change river channel and habitat characteristics, altering the structure and functioning of river systems. Most rivers in the basin are heavily utilized and regulated, with a resultant compromise on the quality and quantity of water in the basin. Flow alterations adversely affect river ecosystems, compromising the resultant goods and services and thereby the dependent human livelihoods as well.

Organisation of the Book

Chapter One introduces the book by presenting background information of major themes and how the book is structured. The chapter highlights ecological changes in the Zambezi River basin using examples from four countries— Malawi, Tanzania, Zambia and Zimbabwe, where the five river sub-catchments in the basin are found. The chapter also describes the major drivers and pressures that are pushing ecological changes in the basin. Snippets of what is to come in each chapter then follow and end the chapter.

Chapter Two illustrates the research method used in the book. It describes the conceptual and theoretical framework which led to the research design that addresses the challenges in the basin. It presents the methodological approach used across the book in determining the ecological changes in the Basin. It also introduces and provides the background information of the Zambezi River and its ancillary five sub–catchments that are the focus of this book. In addition, the chapter provides an overview of the comparative research method and our analytical approach to it.

Chapter Three presents trends in ecological changes in east and southern Africa, and uses the strict definition by focusing on ecological systems. It delves into the causes (drivers and pressures) and status, as well as impacts, of the ecological changes in the aquatic ecosystems within the Zambezi River Basin and the region. The chapter presents the implications that the changes have on the provision of ecosystem goods and services and how the changes are affecting the livelihoods of communities within the Basin. Moreover, it presents a case study from the Manyame River Basin where an example of how different communities perceive and interact with the resources within the Zambezi River Basin can be drawn. In response to the issues raised in the chapter, the authors suggest that communities, policy-makers, researchers, civil society organisations and other players should be involved in developing, planning and conservation of, as well as interventions to restore degraded ecological systems in the Zambezi River Basin and the region at large.

Chapter Four investigates the usefulness of biological monitoring in assessing river health and integrity in the east and southern African region. It explores the use of different bio-monitoring tools as well as biological assays (organisms) to infer the health of aquatic ecosystems in the region. The chapter also presents two case studies; that is the use of biological monitoring in the Manyame River (Zimbabwe) and across several river basins in Tanzania which helped develop the Tanzania River Scoring System (TRSS). The chapter ends by recommending that diatoms and macro-invertebrates are complementary to each other in bio-monitoring programmes and that they can be used to profile rivers for pollution. The authors also advocate for the development of calibrated in-country bio-monitoring tools versus that which has been developed specifically for South Africa as a way forward in biological monitoring of rivers for the region. The chapter is pivotal and connects to the major theme of the book and more specifically, objectives 1 and 2 (see Chapter 2), by providing the scenario on one of the most fundamental ways of monitoring the changes in the health and integrity of the lotic systems of the region, among which the Zambezi River is the largest.

Chapter Five provides essential information which are key to hydrology (Chapter Six) and environmental flows (Chapter Seven). The chapter explains the climate change of five sub-catchments within the Zambezi River basin. It reveals the trends of temperature and rainfall over a period of thirty years. It also projects the climate conditions within the basin for the mid (2050s) and end of the century (2100). Delving into how changes in climatic condition may affect the provision of ecosystem services, the chapter presents community perceptions on this issue. It explains how indigenous knowledge systems and practices help local communities adapt to changes in climatic conditions. This provides an interesting evaluation of real-life practices and concepts against real-climate data. This comparison offers an opportunity to bridge the basic science with applied local knowledge, which is often missing in other studies that treat these ideas separately. The chapter 'ends' by showing how integration of LIKSP in weather forecast can help local people to adapt to climate change. It also shows the role of implementing policy and planning climate related issues in the basin.

Chapter Six focuses on the hydrology of the Zambezi River. In this chapter, considerations of hydrological components such as rainfall, evaporation, stream flow and water balance are analysed for six river sub-catchments of the Zambezi. These are the Lake Malawi, and the Manyame, Ruhuhu, Shire, Songwe and Upper Zambezi Rivers. The analyses show that there are a number of regional modulators of rainfall in the Zambezi Basin which are: the El Niño Southern Oscillation (ENSO), the Inter Tropical Convergence Zone (ITCZ), and Zaire Air Boundary (ZAB) as well as Tropical Cyclones. In addition, a number of changes in the hydrology of the Zambezi River tribu- taries are noted. These are attributed to several factors including climate change (see Chapter 5) and extreme weather events which are made worse by catchment degradation—fuelled by the expansion of agricultural production and demands for fuel wood. The main conclusion of this chapter is that the hydrological trends within the sub–catchments of the Zambezi under study are heterogeneous and therefore there is a need to adopt integrated water resources management (IWRM) practices in order to effectively manage the waters of the ZRB.

Chapter Seven examines the similarities and differences, strengths and limitations of environmental flows (EF) through a review of studies conducted across the sub-basins within the Zambezi River Basin. This is important given the ecological changes, biological monitoring, climate and hydrological trends discussed in the preceding chapters. The chapter examines environmental flow assessments that have been conducted in the riparian countries of the Zambezi River Basin, the methodologies used, policies and legal framework governing the environmental flows and implications of environmental flows on regional integrated water resources management and development planning. The chapter also showcases environmental flow assessments of two case studies; one on the Manyame River in Zimbabwe and the other on the Greater Ruaha Catchment in Tanzania. The authors posit that environmental flows are important in providing the link for integrating water resources development and management to meet the needs of people, agriculture, industry, energy and ecosystems within the limits of available supply and under a changing climate. A brief review of lessons learned from previous EF studies, which can be used to improve future EF assessments and ensure sustainability and continued support of ecosystem goods and services in the basin is provided. This serves to provide, and stresses the need for, information to riparian countries that can help to link between basin development and ecosystem health in order to ensure sustainable development and management of the precious basin water resources.

The last chapter is the Synthesis and, therefore, the concluding chapter of the book. This chapter provides the overviews or 'bigger picture' messages that have emerged from the entire project and preceding chapters. It also tries to address all the objectives and research questions posed. The chapter reiterates the lessons learned, the most outstanding points when it comes to the issues presented before and possible solutions in the context of sustainable use of the resources within the Zambezi River Basin. The chapter also underlines the key messages from the research work.

We hope that this book will fill the knowledge gaps in the basin and stimulate debate around the themes covered to sharpen our understanding and offer solutions to the ecological challenges facing the Zambezi River system.

References

African Development Bank, 2008, 'Clean Energy Investments Framework for Africa', in African Development Bank, ed., Role of the African Development Bank Group. Tunis, 22 pp.

Arnell, N. W., 1999, 'Climate change and global water resources', *Global environmental Change*, 9, 31-50.

Arnell, N. W., 2004, 'Climate change and global water resources: SRES emissions and socio-economic scenarios', *Global Environmental Change*, 14, 31-52.

Bates, B. C., Kundzewicz, Z. W., Wu, S. & Palutikof, J. P., eds, 2008, Climate change and water. Technical Paper of the Intergovernmental Panel on Climate Change, Geneva: IPCC Secretariat.

BBC News, 2015, 'Malawi Country Profile', www.bbc.co.uk/news/world-africa-13864367. 16 June 2015.

Beck, L. & Bernauer., T., 2010, 'Water Scenarios for the Zambezi River Basin, 2000-2050', *Global Environmental Change*, 1-29.

Beilfuss, R. & Dos Santos, D., 2001, 'Patterns of Hydrological Change in the Zambezi Delta, Mozambique', Working Paper# 2. Programme for the Sustainable Management of the Cahora Bassa Dam and the Lower Zambezi Valley. Baraboo, Wisconsin: International Crane Foundation. http://www.savingcranes. org/images/stories/pdf/ conservation/ Zambezi_hydrology_Working_Paper2.pdf.

Brown, D., Chanakira, R., Chatiza, K., Dhliwayo, M., Dodman, D., Masiiwa, M., Muchadenyika, D., Mugabe, P. & Zvigadza, S., 2012, 'Climate change impacts, vulnerability and adaptation in Zimbabwe', IIED Climate Change Working Paper No. 3.

Chagutah, T., 2010, Climate Change Vulnerability and Preparedness in Southern Africa: Zimbabwe Country Report, Cape Town, South Africa: Heinrich Boell Stiftung.

Chenje, M., 2000, State of the environment Zambezi Basin 2000, Harare, Zimbabwe: IUCN, SADC.

Chigwada, J., 2005. 'Climate Proofing Infrastructure and Diversifying Livelihoods in Zimbabwe', IDS *bulletin*, 36, 103-116.

Dagne, T., 2011, 'The Republic of South Sudan: Opportunities and Challenges for Africa's Newest Country', CRS Report for Congress: Prepared for Members and Committees of Congress. Congressional Research Service, 24 pp.

Darwall, W. R. T., Smith, K. G., Tweddle, D. & Skelton, P., 2009, 'The status and distribution of freshwater biodiversity in southern Africa', Grahamstown, South Africa: IUCN, Gland, Switzerland and SAIAB (South African Institute for Aquatic Biodiversity).

Dube, O. & Chimbari, M., 2009, 'Documentation of research on climate change and human health in southern Africa', in Centre for Health Research and Development, ed., Research Report produced on behalf of DBL-Centre for Health Research and Development, Denmark: University of Copenhagen.

FAO/WFP, Food and Agriculture Organization / World Food Programme, 2008, 'Special Report Crop and Food Supply Assessment Mission to Zimbabwe 18 June 2008' [Online]. Available from: http://www.fao.org/docrep/010/ai469e/ai469e00.htm [Accessed 28 July 2016].

Goza, G. O. T. R. O. Z., 2012, 'National Report-Zambia', in Nations, U., ed., The United Nations Conference on Sustainable Development (Rio+20), Rio, Brazil: United Nations.

IHE Delft, 2002, 'Virtual water trade: Proceedings of the International Expert Meeting on Virtual Water Trade', in A. Y. Hoekstra, ed., Research Report Series No. 12. The Netherlands, IHE Delft.

IPCC, 2007, 'Impacts, Adaptation and Vulnerability', in M. L. Parry, O. F. Canziani, J. P. Palutikof, P. J. Van Der Linden, & C. E. Hanson, eds, Contribution of Working Group II to the Fourth Assessment Report of the Intergovernmental Panel on Climate Change, Cambridge, UK: Cambridge University Press, 976 pp.

IPCC, 2014, 'Climate Change 2014: Synthesis Report. Contribution of Working Groups I, II and III to the Fifth Assessment Report of the Intergovernmental Panel on Climate Change' [Core Writing Team, R. K. Pachauri, L. A. Meyer eds], IPCC, Geneva: Switzerland, 151 pp.

Kusangaya, S., Warburton, M. L., Van Garderen, E. A. & Jewitt, G. P. W., 2014, 'Impacts of climate change on water resources in southern Africa: A review', *Physics and Chemistry of the Earth, Parts A/B/C*, 67, 47-54.

Liu, B. & Mwanza, F. M., 2014, 'Towards Sustainable Tourism Development in Zambia: Advancing Tourism Planning and Natural Resource Management in Livingstone (Mosi-oa-Tunya) Area', *Journal of Service Science and Management*, 7, 30-45.

Manyeruke, C., Hamauswa, S. & Mhandara, L., 2013, The Effects of Climate Change and Variability on Food Security in Zimbabwe: A Socio-Economic and Political Analysis', *International Journal of Humanities and Social Science*, 3, 270-286.

Millennium Ecosystem assessment, 2003, Conceptual Framework. Washington D.C: Island Press.

Molden, D., 2007, 'Water for Food, Water for Life: A Comprehensive Assessment of Water Management in Agriculture', London: Earthscan, and Colombo: International Water Management Institute, 118 pp.

MTENR, M. O. T., 'Environment and Natural Resources 2007. Formulation of the National Adaptation Plan of Action (NAPA) on Climate Change', Final Report. Lusaka, Zambia: Ministry of Tourism Environment and Natural Resources.

MTENR, M. O. T., Environment and Natural Resources, 2011, 'The Economics of Climate Change in Zambia', Lusaka, Zambia: Ministry of Tourism Environment and Natural Resources.

Mwendera, E., 2010, 'Situation Analysis for Water and Wetlands Sector in Eastern and Southern Africa', Nairobi, Kenya: IUCN.

Ndebele-Murisa, M. R., Musil, C. F. & Raitt, L. 2010, 'A review of phytoplankton dynamics in tropical African lakes', *South African Journal of Science*, 106, 13-18.

Nel, J. L., Driver, A., Strydom, W., Maherry, A., Petersen, C., Hill, L., Roux, D. J., Nienaber, S., Van Deventer, H., Swartz, E. & Smith-Adao, 2011, Atlas of Freshwater Ecosystem Priority Areas in South Africa: Maps to support sustainable development of water resources. WRC Report No.TT 500/11. Pretoria, South Africa: Water Resources Commission.

Nyamadzawo, G., Gwenzi, W., Kanda, A., Kundhlande, A. & Masona, C., 2013, 'Understanding the causes, socio-economic and environmental impacts, and management of veld fires in tropical Zimbabwe', *Fire Science Reviews*, 2, 1-13.

Rummukainen, M., 2010, 'State-Of-The-Art with Regional Climate Models', Wiley Interdisciplinary Reviews: *Climate Change*, 1, 82-96.

Samuelsson, P., Jones, C. G., Willen, U., Ullerstig, A., Gollvik, S., Hansson, U., Jansson, C., Kjellstrom, E., Nikulin, G. & Wyser, K., 2011, 'The Rossby Centre Regional Climate model RCA3: model description and performance', *Tellus A*, 63, 4-23.

SARDC, SADC, ZAMCOM, Grid-Arendal & UNEP, 2012, 'Zambezi River Basin Atlas of the Changing Environment', Gaborone, Harare and Arendal: SADC, SARDC, ZAMCOM, GRID-Arendal, UNEP.

Schulze, R. E., ed., 2005, 'Climate Change and Water Resources in southern Africa: Studies on Scenarios, Impacts, Vulnerabilities and Adaptation, Pretoria', South Africa: Water Research Commission.

Shoko, D. S. M., 2002, 'Small-scale mining and alluvial gold panning within the Zambezi Basin: an ecological time bomb and tinderbox for future conflicts among riparian states', Commons in an Age of Globalization, the Ninth Conference of the International Association for the Study of Common Property, June 17-21 2002, Victoria Falls, Zimbabwe.

Shongwe, M., Van Oldenborgh, G., Van Den Hurk, B., De Boer, B., Coelho, C. & Van Aalst, M., 2009, 'Projected changes in mean and extreme precipitation in Africa under global warming, Part I', *Southern Africa Journal of Climate*, 22,3819-3837.

Shongwe, M., Van Oldenborgh, G., Van Den Hurk, B. & Van Aalst, M., 2011, 'Projected changes in mean and extreme precipitation in Africa under global warming, Part II: East Africa', *Journal of Climate*, 24, 3718-3733.

Sun, T., Xu, J. & Yang, Z. F., 2012, 'Ecological adaptation as an important factor in environmental flow assessments based on an integrated multi-objective method', *Hydrology and Earth System Sciences Discussions*, 9, 6753-6780.

The World Bank, 2015, Zimbabwe, http://data.worldbank.org/country/zimbabwe [Accessed 28 July 2016].

The World Factbook, 2010, Malawi, https://www.cia.gov/library/publications/the-world-factbook/geos/print/country/countrypdf_mi.pdf. [Accessed 28 July 2016].

Tumbare, M. J. & Zambezi River Authority. The Zambezi River: Its Threats and opportunities. 7th River Symposium, 2004, 1-3pp.

Tweddle, D., Cowx, I. G., Peel, R. A. & Weyl, O. L., 2015, 'Challenges in Fisheries Management In The Zambezi, one of the great rivers of Africa', *Fisheries Management and Ecology*, 22, 99-111.

Unganai, L. S., 1996, 'Historic and future climatic change in Zimbabwe', *Climate Research*, 6, 137-145.

United Republic of Tanzania-URT, 2001, 'National report on the implementation of the convention on biological diversity', Dar es Salaam, Tanzania: Division of Environment, Vice President's Office.

Vörösmarty, C. J., Douglas, E. M., Green, P. A. & Revenga, C., 2005, 'Geospatial indicators of emerging water stress: an application to Africa', AMBIO: *A journal of the Human Environment*, 34, 230-236.

Vörösmarty, C. J., Green, P., Salisbury, J. & Lammers, R. B., 2000, 'Global water resources: vulnerability from climate change and population growth', *Science*, 289, 284-288.

2

A Review of
the Comparative Research Method

Mzime Ndebele-Murisa, Ismael Aaron Kimirei,
Chipo Plaxedes Mubaya, Geoffrey Chavula, Tongayi Mwedzi,
Tendayi Mutimukuru-Maravanyika and Masumbuko Semba

Introduction

The Zambezi River receives its waters from eight countries, namely Angola (18.3 per cent), Botswana (2.8 per cent), Malawi (7.7 per cent), Mozambique (11.4 per cent), Namibia (1.2 per cent), Tanzania (2 per cent), Zambia (40.7 per cent), and Zimbabwe (15.9 per cent) (Figure 2.1). With a length of about 2,600 km and a drainage basin of about 1.3 million km², the Zambezi River is the fourth largest river basin in Africa after the Congo, Nile, and Niger River basins. The basin's ecological changes; physical, chemical and biological health explored via biological monitoring, analyses of climate, hydrology, flow regimes and communities are detailed in Chapters 3, 4, 5, 6, 7 and 8 respectively.

Population growth has taken a dramatic turn in most African countries, with Africa's growth rate of 1.4 per cent, translating to an addition of about 8 million people annually. The continent's population, which was estimated to be 1.2 billion in 2015 is projected to increase by 29 per cent to 1.7 billion by 2030 (Ashton et al. 2001; UN 2015). The increasing population, in turn, increases demand for water and other essential goods and services (especially fish, agriculture and hydropower) from such river basins as the Zambezi. While the current water use in the Zambezi is mainly dominated by hydropower production, increased demand for water supply for domestic use and irrigated agriculture may create water-use conflicts. In turn, climate adds pressure that will likely exacerbate the increasing demand for water in and from the basin, especially changes in rainfall and temperature patterns (detailed information on the climate of the study area is provided in Chapter 5).

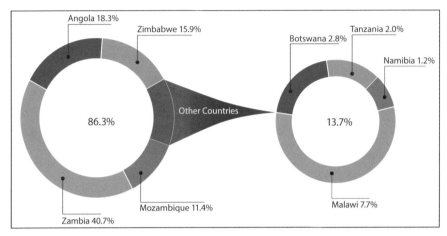

Figure 2.1: Percentage cover of countries comprising the Zambezi River Basin

Hydrological flow regimes and ecological attributes are being altered and impaired by human developments, thereby affecting the ecosystem functions of the Zambezi and many other river basins in Africa. The ecological settings across African water resources, specifically in the Zambezi basin, are changing due to the ever-increasing demand for natural resources, especially water resources, resulting in environmental and other anthropogenic pressures that impact human livelihoods and the catchments' socio-ecological characteristics. Ecological changes and human livelihoods become central in studying and analyzing river flow regimes. However, conducting river-by-river flow and ecological assessments is a challenging task as it requires the mobilization of huge human and non-human resources; and as an alternative, environmental flow analysis for the basin becomes an inevitable endeavor.

Environmental flows (EF), which describe the timing, quantity and quality of water flows required to sustain freshwater ecosystem functioning and human livelihoods, are important for guiding policies and management practices of water resources. Several EF studies have been successfully conducted in the Zambezi River Basin, in almost all riparian countries' sub-basins (Beilfuss and Brown 2010). Each of the eight riparian countries has its own water/river management laws and regulations, which further complicates water resource management in the basin. Moreover, the countries in the basin have varying socio-economic developments and needs for water, which adds an extra dimension to the problem of equitable and sustainable access to and sharing of water resources

(Ashton et al. 2001; Kambole 2003; Odada et al. 2003). Therefore, considering the complexity and importance of the basin, individual river assessments are inevitable. At the same time, regional or basin-wide analysis and management are needed to ensure sustainability and continued support of ecosystem goods and services. An alternative way of accomplishing this is conducting separate river flow and ecological assessments in the sub-basins and comparing findings, and either upscale or use these analyses and assessments as best practices for the unassessed sub-basins or the entire basin. This study attempts to achieve this objective particularly in Chapter 7.

The current chapter documents the research methods and approaches used to assess the different ecological changes that have been occurring in the Zambezi River Basin. We used a mixed approach as well as the Comparative Research Method (CRM) by utilizing both primary and secondary climate, socio-economic, ecological, and hydrological data from sub-catchments within the Zambezi River Basin in Malawi, Tanzania, Zambia and Zimbabwe. The chapter further describes the five sub-basins on which this study and the book are based; provides the conceptual and theoretical framework, and presents the research methods used for data collection and analyses in the different chapters of this book. In particular, the current chapter discusses methods used in analyzing the five sub-catchments in Malawi, Zambia and Tanzania to draw up best practices for use in the Zimbabwe section of the Zambezi River Basin.

A Note on the Comparative Research Method

In the present study, we used the most general Comparative Research Method (CRM) approach, that is, Plausible Rival Hypothesis analysis, which forces the study to examine every potential explanation for any data set (Berman 2002). We also used the emicetic distinction which documents valid principles that describe behaviour in any one culture, considering what the people themselves value as meaningful and important (Brislin 1970). This is eminent in Chapters 3 and 5 which integrate perceptions on climate as well as natural resources and indigenous knowledge systems and practices with climate data analyses. However, we also recognized the limitations presented by CRM in generalizing across cultures and different study areas (Pike 1967; Weiner et al. 2003). This was eminent in Chapter 4, where an attempt to compare the development of biological monitoring in Tanzania and Zimbabwe was rendered unfeasible due to differences in methodological approaches.

When quantitative data is compared, the assumption is that the data is representative, somewhat like a microcosm, of the groups from which it was sampled. The five sub-catchments (including Lake Malawi) used as case studies in the present study (where quantitative data was available), constitute 38 per cent of the Zambezi River Basin (ZRB) in numbers (5 out of 13 sub-catchments); 24.7 and 6.8 per cent of the ZRB in population and areal extent respectively. Therefore, the five sub-catchments were taken as a representative sample of the ZRB. In contrast, when qualitative researchers make comparisons of data, it is usually from nonrandomized samples. The evidence used for comparisons is not generalizable, but may be transferable (Lincoln and Guba 1985). In other words, it is not easy to extrapolate, generalize and make concrete conclusions from such analyses. The qualitative data compared should be carefully crafted to complete a picture of a phenomenon of interest within each group in order to thoroughly understand the phenomenon prior to making comparison (Alasuutari 1995; Morse 2003). In our case we present qualitative data on two sampled communities within the ZRB (see Chapters 3 and 5) and therefore do not generalize or extrapolate the results of these communities across the ZRB. However, we do make generalizations when reviewing literature based on studies conducted across the region, such as the use of local indigenous knowledge and practices (LIKSP) that are presented in Chapter 5.

With quantitative data, the use of meta-analyses can be employed for comparative research. Meta-analysis is the statistical combination of results from two or more separate studies. More often, such analyses have the potential to mislead seriously, particularly if specific study designs, within study biases, variation across studies, and reporting biases are not carefully considered. In addition, variation across studies (heterogeneity) must be considered, although most Cochrane reviews do not have enough studies to allow the reliable investigation of the reasons for it. Again, this was the case for Chapters 4, 5 and 7 of this book where, though data collected may have been of the same nature, differences in frequency of collection and the heterogeneity across the study areas meant that we ended up reporting patterns and trends separately for each sub-catchment.

In addition, a small number of cases (i.e. five as in the river basins in the present study) in CRMs create a propensity to particularize and view each case as unique, limiting generalization; while at times problems are encountered with averaging, where countries and regions are treated as homogenous units. There is, therefore, a need to take note of internal variations that can be encountered withal when making comparisons. In the case of the present study, some differences, particularly in cultures were expected across the five case studies. Nonetheless, some homogeneity was anticipated as a result of the river basins being tributaries of the one Zambezi River (herein treated as a unit) and therefore bearing some semblance of similarity within that unit particularly in ecology, sources of livelihood and ecosystem goods and services. At the same time, internal variations within each river basin were considered in order to avoid the mistake of treating each basin as a homogenous sample.

Conceptual and Theoretical Framework
Conceptual framework, Theory and Hypothesis

The conceptual framework of the study (Figure 2.2) reflects the major theory and themes that are explored in the study as factors which are associated with ecological changes. The theoretical framework is that physical and environmental factors (climate, morphology, river hydrology, flows and water levels, ecosystem health, goods and services) as well as anthropogenic factors (land use and management) affect livelihoods (agriculture, fisheries and mining). These, in turn are ardently determined by the socio-political-economic context which is inclusive of culture, gender dynamics, local indigenous knowledge systems and practices (LIKSP) and perceptions, policy and legal constitutions (Figure 2.2). However, it is important to note that the depicted theoretical and conceptual framework is not all inclusive and feedbacks are also not highlighted despite the factors being interrelated, interdependent, and interconnected and thus presents a much more complex web of interactions than is presented in Figure 2.2.

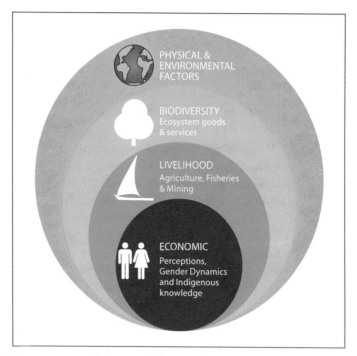

Figure 2.2: Conceptual and theoretical framework

This study also considers the Landscape Approach (LA) which brings the biophysical and socio-ecological issues together to interrogate and find solutions to and tradeoffs between socio-economic development and environmental sustainability in the Zambezi River Basin (Sayer et al. 2013). The LA looks at not only the physical landscape and the people in it but also the institutional conditions (laws, policies, and local customs) that shape how local people use the landscape's natural resources (IUCN 2012). Another definition describes the landscape approach as a tool which is about: "turning an institutional wilderness into an area where everyone agrees about land use, land management and land rights in the different parts of the landscape, and where differing goals are either harmonized or complementary" (Shepherd in IUCN 2012: 7).

Furthermore, the LA encompasses the Integrated Landscape Management (ILM) approach, which is about people and the environment, based on the idea of socio-ecological systems. People's livelihoods are underpinned by services and flow which is built into this coupling of social and ecological systems. We believe that the ILM is the best way to address the poverty and development agenda. In this process, it is important to engage the people whose resilience one seeks to build; while traditional knowledge and local understanding of the issues can contribute in important ways to the solutions that the major international organizations and governments are trying to shape. The importance of this framework is that it enhances the science of studying and the improvement of relationships between ecological processes in the environment and particular ecosystems. This is done within a variety of landscape scales, development spatial patterns, people, and organizational levels of research and policy which makes it perfect in analyzing the heterogeneous and complex landscape of the Zambezi, especially across different borders.

In addition to the ILM, we also employed the concept of landscape ecology. As a highly interdisciplinary science in systems ecology, the landscape ecology concept integrates biophysical and analytical approaches with humanistic and holistic perspectives across the natural and social sciences (Kirchhoff et al. 2012). Landscapes are spatially heterogeneous geographic areas characterized by diverse interacting patches or ecosystems, ranging from relatively natural terrestrial and aquatic systems such as forests, grasslands, rivers and lakes to human-dominated environments, including agricultural, mining and urban settings (Wu 2006). In our case, the ecological landscape was the five river catchments, considered not as fragmented entities bound by imaginary lines (country borders) but as interconnected and interrelated within the heartland of the Zambezi; transcending boundaries and therefore speaking to the higher notions of conservation and sustainable utilization of natural resources. The most salient characteristics of the concept of landscape ecology are its emphasis on the relationships among pattern, process and scale, and its focus on broad-scale ecological, social and environmental issues. This necessitates the coupling of biophysical and socio-economic sciences that lead to the creation of sustainability.

Closely related to the concept of landscape ecology is the ecosystem approach, which provides a framework for the better management of ecosystem services, such as carbon storage and sequestration, freshwater and nutrient cycling, and biodiversity protection that require larger interventions. This approach draws insights from contemporary debates in conservation biology, tied to social-ecological systems, political ecology, agro-ecology, ecoagriculture, and rural and agricultural development, among others. The integration of ecological systems, agroecosystems, livelihoods, and institutions in landscape management, is considered to offer a promising framework for community based socioecological resilience in the context of climate change and other risks.

Study Area

The Zambezi River Basin (located between 8° and 20° South, and 16.5° and 36° East), drains an area of about 1.3 million km². The basin covers about 23.4 per cent of the total area of its riparian countries and is the most shared river in the southern African region. The river flows over a distance of nearly 3,000 km, dropping in altitude from its source in the Kalene Hills of North Western Zambia, 1,585 meters above sea level (masl), to its delta in the Indian Ocean, 200 km north of the Mozambican port of Beira. The river's major tributaries include the Luena and Lungue-Bungo in Angola; the Chobe in Botswana; Shire in Malawi; Luiana in Namibia; Kabompo, Kafue and Luangwa in Zambia; and Gwayi, Manyame and Sanyati in Zimbabwe as well as the Lumbira, Ruhuhu and Songwe in Tanzania. The Cuando or Kwando (Angola), Chobe (Botswana), Barotse and Kafue (Zambia) flood plains are some of the most significant wetlands that form major tributaries of the Zambezi which also play important roles in regulating flooding as well as the flow of the river. The Zambezi River and its dense network of tributaries and associated ecosystems constitute one of southern Africa's most important natural resource systems.

The agroecological zone of the Zambezi Basin is semi-arid with low and erratic rainfall that is received between October and April. There have been reports of increasing frequency and intensity of extreme weather events such as droughts and flooding in this zone. Most rivers in the Zambezi Basin drainage network are ephemeral. Two major dams built along the Zambezi are Kariba and Cabora Bassa. The former straddles Zambia and Zimbabwe and Cabora Bassa Dam lies wholly in Mozambique. Lake Kariba is the third largest man-made lake by size and the largest reservoir by volume in the world. The Zambezi and Sanyati rivers contribute 80 and ten per cent of the water flow into Lake Kariba respectively. Lake systems of note in the Zambezi River Basin are the two aforementioned reservoirs as well as Kafue Dam in Zambia with Lake Malawi being the largest and only natural lake among the major lentic environments of the Zambezi. All the river courses, lakes and wetlands are not only important

as contributors of the water flow and water levels of the Zambezi, but also form
the major source of livelihoods for the riparian communities around them. In
this comparative research network, the Barotse Flood Plain (Zambia), Manyame
River (Zimbabwe), Ruhuhu and Songwe Rivers (Tanzania) and the Shire River
in Malawi were studied.

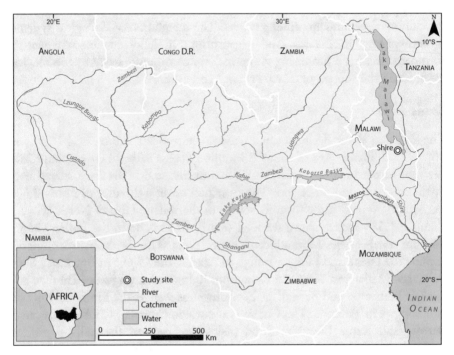

Figure 2.3: The Zambezi River Basin catchment and its major tributaries. The five
sub-catchments under study are highlighted by a double ringed circle

The Barotse Flood Plain System

The Barotse Flood Plain System (BFPS) is part of a bigger political and geographic
system in Zambia. It is located at an elevation of 993 masl. Its central location
is 15°40' South and 23°10' East (Figure 2.4). The flood plain system is located
along parts of the Upper Zambezi River and contributes towards maintaining the
perennial flow of the Zambezi River. The Upper Zambezi River Basin extends
from its source, 25 km south-east of Kalene Hill in Mwinilunga District, north-
western Zambia, through Angola, Barotseland to the Victoria Falls. The basin
comprises a series of broad flood plains that are separated by low sand plateaus
with small and scattered swamps, all of which are situated on a relatively flat
landscape (Timberlake 1998). The Zambezi River has a single peak hydrogram
that results in flooding during the period from December to June. All year

round, due to the rise and fall of water levels, several channels open and close throughout the course of the main river channel, and flood plains and oxbow lakes are common features. The ecological characteristics of the Barotse Flood Plain and human production system that it supports are highly dependent on the timing and duration of flooding (Timberlake 1998). The onset of the annual flooding varies greatly and can occur any time between December and March. Peak flooding occurs in April, after which the flood waters recede from May to July (IUCN 2003).

The Barotse Flood plain system is located in the Western Province of Zambia and is spread over four of the seven administrative districts in the province, namely Mongu, Kalabo, Lukulu and Senanga (Figure 2.4). The total population of the four districts is estimated at 225,000 people (IUCN 2003). The Lozi people occupy the flood plain area that has a dual administrative system consisting of the Government of Zambia and the Barotse Royal Establishment that is ruled by the King or Litunga. The BFPS covers a total area of 1,097,269 ha and comprises six sub-systems, namely Lungue-Bungo (109,000 ha), Kabombo (18,800 ha), Nyengo (71,500 ha), main Barotse flood plains (770,000 ha), Luena (89,669 ha) and the Lueti and Lui Swamps (38,300 ha) (ZAWA 2006).

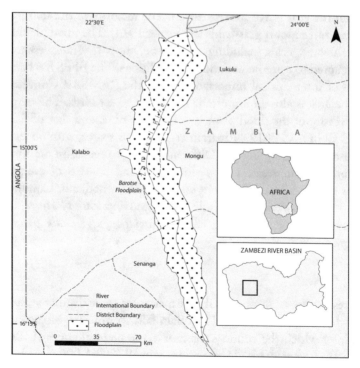

Figure 2.4: Map of Barotse Flood Plain (Zambia) with the Zambezi River flowing through the middle of the flood plain

The Barotse flood plain is a flat plateau that stretches from the confluence of the Zambezi River with Kabongo and Lungwebungu rivers in the north, to a point 230 km above Ngonye Falls, south of Senanga (AAS 2012). It is the second largest wetland in Zambia after the Bangweulu system and because of its vastness; the Barotse Flood Plain system is considered the largest of the Zambezi's wetland complexes (Flint 2009). The flood plain is also considered one of the world's most complex systems due to its intermittent feeder rivers and streams and its flood regulating effects. The flood plain is mostly 30 km wide along its length and reaches 50 km at its widest point north of Mongu.

Physiography and Geomorphology

The BFPS consists of flood plains, dambos and pans. These are important to communities as they provide suitable conditions for the cultivation of rice. The system is dominated by Kalahari sands that cover the surface up to a depth of about 300 m. Floods leave behind fertile grey to black soils overlaying the Kalahari sands. The soil types range from sandy to loam and clay and they remain wet throughout the year, except in areas where they have been artificially drained. Where drained, the surface becomes firm but the bottom soil remains wet throughout the year (CRP in AAS 2012). The majority of the soils in BFPS are not drained and support grasslands (Brammer 1973). The flood plain is home to several herbivore species, including red Lechwe, Hippos, Wildebeest and Tsessebe that are adapted to extreme wet and dry conditions. The BFPS has been declared a wetland of international importance under the RAMSAR Convention (AAS 2012). The area is also an important breeding site for birds. The herpeto-fauna species diversity of the flood plain area is eighty-nine, and that of fish is eighty species (CRP in AAS 2012). Riparian vegetation is sparse, with no trees in areas where seasonal flooding occurs. Different types of vegetation are found in the upland and lowland areas. The vegetation upland consists of evergreen semi-deciduous and deciduous *Brachystegia Julbernadia* woodland. Common woody plant species in the woodland include *Pterocarpus angolensis plurijuga, Guibourtia coleosperma, Afzelia quanzensis, Baikiaea plurijuga, Julbernadia paniculata* and *Brachystegia* species.

Climate

Like the rest of the country, the BFPS experiences three distinct seasons: warm-wet, cool-dry and dry-hot. The flood plain falls within agro-ecological zone IIb of Zambia of which the country has four agroecological zones (I, IIa, IIb and III). Zone IIb is characterized by medium rainfall (800-1,000 mm) but is prone to droughts and flooding. The Barotse Flood Plain is considered vulnerable to climate change. Current climate trends indicate reductions in river flow and velocity including the regularity and intensity of the flood plains (Flint 2007).

The reduced flow is coupled with increased temperature and evaporation. Thus far, climate change has resulted in increased frequency of extreme weather patterns such as droughts, severe storms, higher annual floods, flood failure, increased temperatures (from 1 to 2°C between 1970 and 2004) and more severe winds. Impacts of climate change that are experienced include decrease in crop production due to low rainfall, shifts in rainfall patterns, water shortages and unavailability due to high temperatures and high evaporation rates, and increased occurrence of pests and diseases (Nhamo and Chilonda 2003; Ministry of Finance 2013). Detailed analyses of the climate of the Barotse Flood Plain are provided in Chapter 5 of this book.

Land Use

There is no formal land use planning in Barotse but this has become increasingly necessary due to an increase in population. This increase has led to increased pressure on resources such as cropland, pastures, fisheries and forests as well as land and water uses both in the plains and in the uplands. Although the BFPS has huge potential to support agricultural production, this potential is not realized due to high variability and vulnerability of the system to climate, among other factors (CRP in AAS 2012). The system is characterized by extreme climate events such as excessive and unpredictable floods that destroy crops and agricultural infrastructure. Because of the flooding system, people in the Barotse are transhumant and follow the rising and receding of floods. Their agricultural production is therefore linked to, and follows, human migration patterns.

The main growing season in the flood plain is between November and April and diverse farming systems exist due to different soil types on both the dry and wetlands. Many crops are grown in the plains including the following food and cash crops: maize, rice, cassava, sweet potatoes, mangoes and cashew (CRP in AAS 2012). The soils in the flood plain are most suited for paddy rice production and this dominates agricultural production. Cashew was introduced by a private company in the early 1980s as a cash crop as well as to promote afforestation in Barotse.

Because of the vast grasslands, livestock production is part and parcel of the agricultural production in the flood plains. Cattle rearing dominates livestock production and, over time, the total livestock population of the flood plain has declined due to disease and pasture shortage. Fishing is also a key livelihood activity in the Barotse, and the fishery is important for food security and the livelihoods of about 70,000 people who depend on it. However, due to poor management strategies and several other factors, fish stocks have declined over time, resulting in an urgent call for sustainable management of the fisheries by the Barotse stakeholders. Efforts are currently underway to adopt fisheries co-management, which is already enshrined in the Zambia Fisheries Act of 2011. Aquatic plants found in the BFPS are also central to people's livelihoods and these include papyrus, thatching grass and palms that are used to make ropes and baskets.

The Manyame River Catchment

The Manyame catchment is one of the seven major river basins of the Zimbabwean hydrological water management system. It lies between 18° and 15° South and 28° and 32° East and its source is in Marondera (Figure 2.5). The river eventually drains into the Zambezi River downstream of the Kariba Dam and upstream of the Cabora Bassa Dam to the northern part of Mozambique. The Manyame has a total estimated catchment area of 40,497 km² (Vincent and Thomas 1960; Mugandani et al. 2012). This is the most urbanized catchment in Zimbabwe, covering four administrative provinces, namely Harare Metropolitan, Mashonaland East, Mashonaland West and Mashonaland Central (Figure 2.5). The Manyame River is the main water source for Harare-which, in turn, is the most densely populated province in Zimbabwe (population of 2,123,132 (Zimbabwe National Statistic Agency 2012). However, the river flows mainly within the Mashonaland West Province, which has an area of 57,441 km² and a human population of 1,501,656 (Zimbabwe National Statistic Agency, 2012). This translates into a densely populated province which is divided into 13 districts that are further divided into 222 wards. Only two out of eight other provinces in Zimbabwe (excluding Harare) have higher populations than Mashonaland West Province.

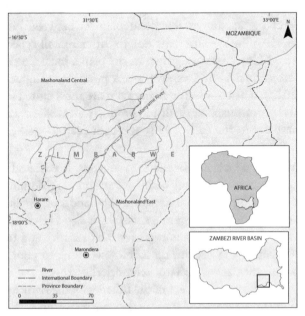

Figure 2.5: Map of the Manyame River Catchment, Zimbabwe

The Manyame River catchment is divided into three main plains based on the spatial distribution of the catchment terrain. The upper part of the catchment consists of areas around Marondera, Chihota and Seke. The middle section consists of the Chinhoyi, Banket, Mukwadzi, Mutorashanga, Raffingora, Mhangura and Guruve areas. The lower section of the Manyame River consists of the low-lying Dande communal areas which are below the Zambezi escarpment covering the Chirundu, Mushumbi, Mahuwe, Muzarabani and Mukumbura areas (Thornton and Nduku 1982).

Physiography and Geomorphology

The Manyame River catchment includes streams draining part of the Great Dyke, which is a linear geological feature that extends nearly north-south through the centre of Zimbabwe passing to the west of the capital city Harare. The Great Dyke consists of a band of short, narrow ridges and hills spanning approximately 550 km. The Great Dyke has large commercial deposits of nickel, copper, cobalt, gold and platinum group metals. The stream beds in some sections of these areas comprise ultramafic rocks strongly enriched in magnesium-bearing minerals (Makore et al. 2012). In the Lower Manyame catchment, the area is composed of highland granite, greenstone terrain and the sediments of the Zambezi Valley (Moore et al. 2009). The greenstones along with andesine and greywacke tuffs are gold-bearing. There are therefore high rates of gold occurrences along the river (Moore et al. 2009).

Climate

The Manyame catchment area has a wet-dry tropical climate with annual rainfall ranging from 450 to 880 mm (Proctor and Cole 1991). The area has mean annual temperatures of $22.5 \pm 5.1°C$ with a mean monthly maximum of $29.5 \pm 6.5°C$ and a mean monthly minimum of $18.9 \pm 5.8°C$ (Meteorological Services Department of Zimbabwe: data from 1965 to 2014). In general, Zimbabwe is experiencing increases in temperature, and decreasing rainfall (Lotz-Sisitka and Urquhart 2014) and the Manyame Catchment is no exception to these trends. Temperatures have increased by 0.4°C in the period 1900 to 2000, with the period from 1980 to date being the warmest (Ministry of Environment and Natural Resources Management, 2013). In the same period, rainfall has declined by five per cent, with the last decade of the century being the driest years (Ministry of Environment and Natural Resources Management 2013, Lotz-Sisitka and Urquhart 2014). Detailed analyses of the historical and projected climate of the Manyame and the other river catchments are presented in Chapter 5 of this book.

Land Use

Streams in the Manyame River catchment area flow through four different land use categories identified following Anderson et al. (1976) and these include; urban, mining, communal and commercial agricultural areas (see Figure 4.1 in Chapter 4). Due to population growth, uncontrolled urbanization and industrialization, and the economic downturn in the past decade, various town councils in the study area do not meet the required standards for sewage treatment, garbage collection and urban drainage. In addition, because the catchment is rich in minerals, mining is a major socio-economic activity. In this regard, the Manyame catchment is prone to anthropogenic activities including urbanization, agriculture, sewage treatment, sand extraction, mining, damming and industrialization, which subsequently can negatively affect river hydrology and biota in its streams. The streams in the study area, therefore, receive pollutants from various point and diffuse sources and their habitats have been greatly altered, resulting in stream health deterioration, eutrophication, organic and metal pollution, among other threats (Utete et al. 2013; Bere and Mangadze 2014).

In communal agricultural areas, a combination of poor agricultural practices (stream bank cultivation, overgrazing, and soil erosion) and high human population densities have negative effects on water quality of streams draining these areas. Commercial agricultural areas are relatively pristine compared to the other three land-use types. These areas are characterized by mature deciduous riparian forest strips which act as riparian buffers, thus protecting water resources from nonpoint sources of pollution and providing bank stability and aquatic habitats. In addition to the four land use categories identified above, large dams are prominent in the upper Manyame catchment and tend to be eutrophic, with the most notable one being Lake Chivero with incidences of localized pollution reported (Moyo 1997; Hranrova et al. 2002; Nhapi et al. 2002; Magadza 2003; Ndebele and Magadza 2006) and some extended pollution from hypereutrophic Lake Chivero more recently observed in downstream reservoirs such as Darwendale and Biri Dams (Regean Mudziwapasi, Personal communication).

The Shire River Catchment

The Shire River is located entirely in Malawi. Its basin lies between latitudes 14°20' and 17°08' South and longitudes 34°15' and 35°33' East (Figure 2.6). The river is divided into three main sections or reaches, i.e. Upper Shire, Middle Shire and Lower Shire. The Shire's source is Lake Malawi, upstream of Mangochi, at an elevation of 474 masl. The river passes through Lake Malombe, then flows in a south and south-easterly direction till upstream of Nsanje (Water Department/ UNDP 1986). From there, it then flows further south until it joins the Zambezi in Mozambique. The first reach of the Shire River extends from Mangochi to Matope over which distance the river loses an elevation of 17 m. The Middle Shire extends from Matope to Maganga which is a few kilometres downstream of Kapichira Falls. In this reach, the Shire passes through a series of rapids, including Nkula Falls, Tedzani Falls, Mpatamanga Falls and Kapichira Falls. The Shire River has a total fall of 366 m in the section from Matope to Maganga. And in the last reach of 213 km from Maganga to its confluence with the Zambezi, the Shire flows over alluvial formation, and in an area with a flat topography. The total catchment area of the Shire in Malawi is 18,945 km² excluding Lake Malombe (303 km²).

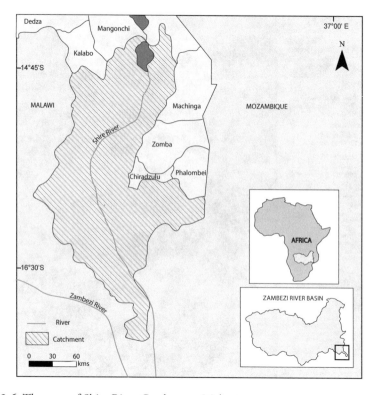

Figure 2.6: The map of Shire River Catchment, Malawi

Physiography and Geomorphology

Out of the total area of 18,945 km² of the Shire catchment, nearly 72 per cent lies between 200-1000 m elevation, about 20 per cent is below 200 m, and about 8 per cent is above 1,000 masl. The Elephant and Ndindi Marshes constitute low lying areas. Along the south-western boundary covering the Matundwe range, and the Natundu hills west of Nsanje, the formation comprises sedimentary and basalt rocks. The Shire River flows mostly along the fault-through of the East African Great Rift Valley System, which extends from Ethiopia down to the Zambezi River in Mozambique.

Soils on either side of the Shire and the lower part of the river south of Chikwawa are mainly calcimorphic alluvial (The soil types in Malawi, Figure 2.7). The soil types are classified into the following types: sand, sandy loam, sandy clay, clay loam, clay, fluvic and leptosol. The soils in the Shire Basin are dominated by sandy clay (71 per cent) with low infiltration capacity followed by

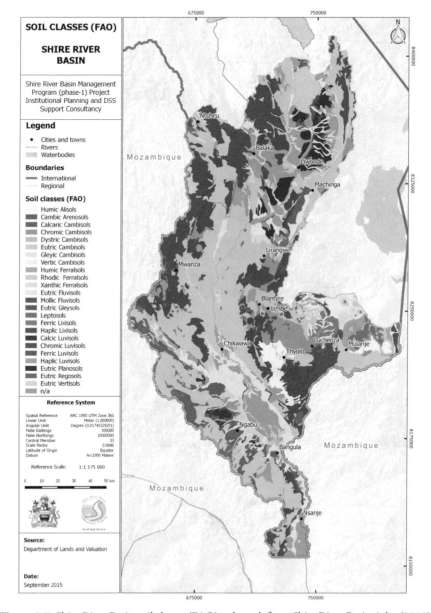

Figure 2.7: Shire River Basin soil classes (FAO), adapted from Shire River Basin Atlas (2012)

sandy loam (17 per cent) (Jessen and von Christierson 2015). The upper part of the Shire is underlain by alluvium; the middle reach is dominated by Pre-Cambrian Basement Complex rocks, and the lower areas again by Quaternary Alluvium. The middle reach is dominated by ferruginous and lithosols, whereas the Elephant Marsh and other low-lying areas are dominated by hydromorphic soils with mostly Karoo sedimentary rocks outcrop on the south west side of the Lower Shire Valley which lie on or are faulted against the underlying Basement Complex. The basal beds of the succession are conglomerates and sandstones, which are overlain by sequences of sandstones, mudstones, shales and coal seams. The upper sandstones and marls become increasingly red in colour. The Karoo sediments are well cemented by calcite and indurate; the primary porosities are thus low, and any more permeable horizons are related to secondary fracturing or enlargement along the well-developed bedding planes. Karoo sediments probably underlie much of the Lower Shire alluvium (Water Department/UNDP 1986).

The Quaternary alluvium, characterized by deposits of colluvial, fluviatile and lacustrine, is well developed along the shores of Lake Malawi and also around Lake Malombe and Lake Chilwa. There are also extensive alluvial deposits in the upper and lower tracts of the Shire River Valley and in the Bwanje Valley. The deposits are unconsolidated and have been formed by deposits from rivers debouching from the rift escarpment and along the lakeshores from lacustrine sedimentation. The sedimentary succession comprises highly variable sequences of clays, silts, sands, and occasional gravels. The mineralogy of rock fragments in the alluvium show that the deposits have been derived from the strata at the sides of the Rift Valley.

As mentioned before, the Shire Basin is part of the East African rift system. The rift consists of eastern and western branches which dissect the entire eastern part of Africa. The Malawi rift, which is a southern extension of the western branch of the Rift System, extends 900 km from Rungwe volcanism in Tanzania to the Urmegraben in Mozambique. In Malawi, it is a single linear zone of extensive "*en echelon*" down faulting, occupied by Lake Malawi, Lake Malombe, and the River Shire. Metamorphic and Crystalline igneous rocks form most of the basement in Africa, and underlie much of Malawi. The geology around the Malawi rift is dominated by Basement Complex gneisses and granulites. Overlying the basement are limited Permo-Triassic Karroo sequences and Cretaceous red beds in the north and south, tertiary lacustrine sediments along the lake shore, Shire River and lake beds. There are igneous rocks and dykes and sills among the sedimentary rocks. The Quaternary Alluvium has been affected by faulting. The post Basement Complex sedimentary sequences were probably deposited in a series of tectonically controlled basins, which have been affected by subsequent warping, faulting and erosion. The most recent Quaternary alluvial material has been deposited in the bottom of the rift valley and lies on older strata (Water Department/UNDP 1986).

Climate

The climate around the Shire River Basin is tropical continental in nature with three main seasons: a hot wet season from November to April; a cold dry season from May to September, and a hot dry season from October to mid-November. Three synoptic systems bring rainfall to the basin: The Inter-Tropical Convergence Zone (ITCZ), the Zaire Air Boundary (ZAB) or the Congo Air Mass, and Tropical Cyclones as they veer away from the normal east to west path in the Mozambique Channel. Rainfall ranges from as low as 500 mm in the extreme south and south-western part of the basin to more than 1,400 mm in the Upper Zambezi and Kabompo sub-basins, in the north-eastern shores of Lake Malawi in Tanzania, and in the southern border area between Malawi and Mozambique. In the lower course of the river in Mozambique the influence of the summer monsoon increases the levels of precipitation and humidity. Temperatures are also higher determined more by the latitude and less by altitude as the river descends from the plateau. The upper and middle course of the river is on an upland plateau, and temperatures, modified by altitude, are relatively mild, generally between 18-30°C. The winter months (May to July) are cool and dry, with temperatures averaging 20°C. From August to October, there is a considerable rise in average temperatures, particularly in the river valley itself where the values may be as high as 40°C. The rainy season lasts from November to April.

Land Use

A detailed account about land use practices in the Shire River Basin was presented by Nanthambwe (2013). A common factor across Malawi and the Shire River Basin is the predominance of smallholder subsistence farming on customary land tenure based on low inputs, poor conservation practices and extensive system of livestock farming with its long record of concerns of associated environmental sustainability. In the forest areas, encroachment of agriculture, tree cutting for fuel wood and charcoal often in environmentally fragile areas are a common and concerning practice.

The Shire River Basin is the largest Water Resources Area in Malawi. It covers 18,945 km² and represents about 16 per cent of the country's total geographical area. The basin is home to approximately 5.5 million people, the majority of whom live in the rural areas where they depend on the natural resource base for the sustenance of their livelihoods. Both government and the private sector have invested heavily in the Shire River Basin for social and economic development and growth of the country. Among these developments are the Kamuzu Barrage at Liwonde, constructed in 1965 to regulate the flow of the Shire River; the Walkers Ferry near Nkula by the Blantyre Water Board from where much of the water supply for the city of Blantyre and its conurbations is abstracted, and the Nkula A and B, Tedzani and Kapichira hydroelectric power stations by the Electricity Supply Corporation of Malawi (ESCOM) from where almost 98 per cent of the country's hydroelectric power is generated. In addition, other developments on the Shire River include the Illovo Sugarcane plantations in the Lower Shire by the Illovo Group, which is one of the largest sugar cane plantations in the country (Nanthambwe 2013) as well as numerous agricultural and non-agricultural projects. These investments offer employment to many Malawians and substantially contribute to the country's gross domestic product (GDP).

The Shire River Basin is also an arena of diverse activities by other players, particularly non-governmental organizations (NGOs) that are involved in projects that aim at improving people's livelihoods. In tandem with this drive, international development agencies are also engaged in similar activities that complement government efforts to improve the welfare of the people within the basin (Nanthambwe 2013). The Shire River Basin is traversed by a dense network of river systems, the major ones being the Nkasi, the Rivirivi, Lisungwi, Wankulumadzi, Likabula, Mwanza, Mwamphanzi, Thangadzi East, Thangadzi West and others, some of which are been exploited for irrigation particularly in the lower reaches of the basin. In addition, the basin is also home to wildlife protected areas that include, the Liwonde National Park, Lengwe National Park, Majete Game Reserve, and important but fragile ecosystems such as the Elephant and Ndindi Marsh. All of these protected areas possess very rich biodiversity of flora and fauna.

Due to the extreme dependence on the natural resource base by the communities living in the basin, the catchment of the Shire is under intense environmental pressure that has translated into severe land degradation in some parts of the basin such as in the Upper Lisungwi, Upper Wankulumadzi and the Blantyre Escarpment. Although these areas exhibit glaring land degradation, much of the various micro-systems within the basin are equally under threat from poor resource management. With the emergence of climate change, the basin will experience further degradation, which would not only impact adversely on the communities that live in the basin but also on the overall social and economic development and growth of the country.

The Ruhuhu and Songwe River Catchments

The Ruhuhu and Songwe River catchments in Tanzania form part of the Lake Nyasa/Malawi Basin. This part of Lake Nyasa/Malawi Basin is located in the south-western part of the United Republic of Tanzania, between 8°57' and 11°36' South and 32°47' and 35°59' East (Figure 2.8) (Nindi 2007; MoW 2013a). The Lake Nyasa/Malawi Basin forms the north-eastern most portion of the Zambezi River Basin. The Lake Nyasa Basin is trans-boundary, with Malawi owning the largest portion of the total catchment (64 per cent), followed by Tanzania (28 per cent) and Mozambique (8 per cent) (Bootsma and Jórgensen 2004; MoW 2013b). A total of ten rivers from the Tanzanian side flow into Lake Nyasa/ Malawi. The Ruhuhu (14,211 km²), Songwe (2,490 km²) and Lumbira (2,153 km²) rivers are by far the largest, with Ruhuhu occupying over 50 per cent of the entire Lake Nyasa/Malawi catchment with a total drainage area of about 29,600 km² (Kingdon et al. 1999; MoW 2013a; LNBWB 2014). The Ruhuhu and Songwe rivers, and many other small rivers, drain into Lake Nyasa, which discharges into the Shire River in Malawi. Shire then pours into the Zambezi River in Mozambique. The Tanzanian catchment provides about 53 per cent of total inflow into Lake Malawi/Nyasa (Chiuta and Johnson 2010; World Bank 2002). The Ruhuhu River catchment contributes about 20 per cent of the annual flows into Lake Malawi/Nyasa (Kidd 1983), while the Songwe and Kiwira rivers contribute only about five and four per cent respectively (Branchu et al. 2010). Population trends are not available for the Ruhuhu and Songwe river basins specifically, but from the districts in which these rivers originate. While Ruhuhu River Basin has the larger catchment area compared to Songwe, the former is moderately populated (133,218) and the latter has approximately 560,647 people, and is therefore densely populated and highly cultivated (Bootsma 2006; URT 2013).

Physiography and Geomorphology

The Lake Malawi/Nyasa Basin is characterized by mountainous and hilly landscapes, with plateau being the most common land form. The Rungwe and Kipengere mountains are over 2,000 masl, and are the highest features, while Lake Malawi/Nyasa, which is located at 470 masl, is the lowest in altitude. Lake Malawi/ Nyasa is located within the western arm of the East African Rift Valley system. The Livingstone Mountains are also a significant feature, which separates the Ruhuhu River from the Great Ruhuhu River system. The Songwe River originates from the Rungwe Mountains between 2,000 and 2,400 masl, navigating through plateaus, escarpments (1,100-1,300 masl) and the lakeshore plains (500-1,000 masl) to eventually pour into Lake Malawi/Nyasa (Chafota 2012). The lower section of the river is dominated by a flood plain, which receives sediments from upstream. Due to the frequency of flooding, which is caused by poor land use in the plateaus and escarpment sections of the river, the Songwe River changes its course every two years (Chafota 2012). The Ruhuhu River emanates from the Kipengere Range which is also known as the Livingstone Mountains (about 2,000 masl). The mountains are located south of Njombe and the river flows into Lake Malawi/ Nyasa at a distance of about 160 km from the source.

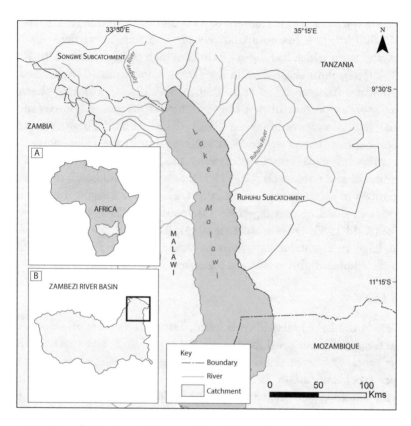

Figure 2.8: Map of Ruhuhu and Songwe River Sub-catchments in Tanzania

The climate of the Ruhuhu and Songwe rivers is broadly characterized by wet (November to May) and dry seasons (June to October). While the wet season is characterized by high temperatures and rainfall levels, the opposite is true for the dry season. Air temperature varies seasonally and between the catchments, with Ruhuhu having slightly higher temperatures than Songwe catchment. Ruhuhu and Songwe have long-term mean annual temperatures of 19.1±8°C and 17.6±6.5°C, with a mean monthly maximum of 27.2±1.9°C and 24.2±1.5°C; and a mean monthly minimum of 16±2.7°C and 11.1±3.1°C, respectively (Tanzania Meteorological Agency, data from 1980 to 2014). Temperatures vary from 16°C and 20°C during the rainy season and between 14°C and 18°C during dry season for Songwe River catchment, and between 20°C and 24°C for the rainy season and 18°C and 21°C during the dry season for Ruhuhu River catchment. According to the Tanzania Meteorological Agency (data from 1980 to 2014), Ruhuhu and Songwe catchments on average receive a long-term total annual rainfall of 1,048 and 903 mm respectively. The long wet season receives more rainfall compared to the short dry season (MoW 2013a).

Despite the fact that the Ruhuhu River catchment is the largest in size (more than 14,000 km²), it has comparatively lower precipitation than the Mbaka River Catchment which has about 20 times as high precipitation as the Ruhuhu (MoW 2013a). Rainfall prediction for Tanzania indicates an increase of 45 to 50 mm year-¹ between 2010 and 2039 (MoW 2013a). Climate change and variability are important in that they affect precipitation, evapo-transpiration and the rainfall-evaporation ratio (R:E) (Kidd 1983). These are therefore significant direct drivers of ecosystem change within the Lake Malawi/Nyasa Basin which influence rainfall availability and variability, rainfall-evaporation ratio and lake and river levels (Crul 1997). Lake Nyasa is known to be sensitive to variations in rainfall-evaporation ratio (R:E) where a slight increase in R:E can result in flooding events, while a decrease may cause zero outflow from the lake (Kidd 1983; Nicholson et al. 2014). Annual variations in rainfall have caused large inter-annual variations in water levels in Lake Malawi/Nyasa, which can modulate fish speciation processes (Crul 1997).

Land Use

Land-use in the Lake Malawi/Nyasa Basin, Tanzania, varies from woodland and natural forest (35.4 per cent), bush lands (31.9 per cent), area under cultivation (17 per cent), grasslands (14.6 per cent), permanent swamps and water bodies (1.1 per cent), and urban areas (0.1 per cent) (MoW 2013a). Agriculture is the main economic activity, whereby narrow elongated plots of cassava, paddy, groundnuts, maize, and other minor crops are cultivated (Nindi 2007). The average land cover for agriculture is estimated to be only 17 per cent in the Lake Nyasa catchment (MoW 2013a). In contrast, the Songwe River catchment cultivated land accounts for about 74 per cent of the total land area (Chafota 2012). Most farmers practice rain-fed agriculture; however, some irrigation projects are taking shape in the basin, where an estimated 3,000 ha of land is currently under irrigated agriculture, mostly paddy (Nindi 2007; MoW 2013b). There are several mining projects, which are both small and large scale in nature in the catchments: with some at the exploration and others at the mining stages (Kreuser 1991). These range from alluvial gold mining in the upper reaches of the Ruhuhu River to coal and chrome mining and prospecting in the Songwe, Mchuchuma and Kiwira catchments (Ashton et al. 2001; MoW 2013c). The active deforestation in the Livingstone Mountains and Matengo Highlands, which is caused by agriculture and mining prospection, result in increased sedimentation in the rivers and consequently in Lake Nyasa/Malawi (Nindi 2007).

The effects of deforestation and the resultant soil erosion can be interpreted from increasing sediment. There are also plans to develop hydropower projects in the Rumakali and Songwe rivers. This may stabilize the river bed and increase dry season flows; but could potentially adversely affect the migration of the potamodromous fish species, especially *Opsaridium microlepis* which is already threatened by river health deterioration in the upper and middle sections of the rivers (NBI 2007; Chale 2010; SMEC 2013). Deforestation and consequent increase in turbidity threaten the International Union for Conservation of Nature (IUCN) red listed fish species *O. microlepis* and *O. microcephalus* (IUCN 2006; Darwall 2008) which are economically important fish species in the Lake Nyasa Basin (Turner 1995; Nindi 2007; Chale 2010; Chafota 2012) and are listed as 'endangered' and 'vulnerable' respectively, under the IUCN red list of endangered species (IUCN 2015). Increase in agriculture production creates a potential threat for increase in nutrient levels and pollutants which can result in a decrease in water quality and increased levels in indicative parameters of pollution such as turbidity. The aforementioned fish species are threatened by turbidity (Duponchelle et al. 2000; Chale 2011) which is caused by soil erosion and deforestation in the catchments. It is, however, important to note that literature refers to turbidity affecting fish in Lake Nyasa but actual measurements of turbidity are elusive. However, in such cases, transparency measures are often used as a proxy (Chale 2011).

The Study Approach

A comparative analysis of five river catchments within the Zambezi Basin, namely the Barotse Flood Plain (Zambia), Manyame River (Zimbabwe), the Shire River (Malawi), and the Ruhuhu and Songwe rivers in Tanzania was conducted. Although they represent two separate river systems, the Ruhuhu and Songwe rivers are amalgamated into one section and represent the portion of the Zambezi Basin in Tanzania, which pours into the Lake Nyasa/Malawi Basin. These two river systems and the Shire River drain into and out of Lake Nyasa/ Malawi respectively with the Shire eventually pouring into the Zambezi River in Mozambique. Therefore, reference is advertently made to the Lake Nyasa/ Malawi Basin which sits intricately between the Ruhuhu/Songwe and the Shire River catchments as a receiver of waters of the former (Ruhuhu and Songwe are its tributaries) and outlet of the latter (Shire) respectively. Malawi, Tanzania and Zambia have conducted river assessments in the major rivers that have been reputed as successful in advocating for coping strategies and innovations that communities can pursue in the wake of environmental changes. We therefore combined primary data (from the Manyame River catchment in Zimbabwe) with secondary data analyses (from the respective river catchments in Malawi, Tanzania and Zambia) in order to compare the river catchments as well as to establish best practices. This learning approach means that the 'seemingly successful' models in the other river catchments were used as a basis of comparison to analyse the vulnerability and adaptation capacity of communities in the Manyame Catchment. This then helped ascertain whether these vulnerabilities, adaptations and best practices were universal or applicable across the four river catchments within the Zambezi River Basin.

In addition, our research investigated communities' perceptions and their current coping and/or adaptation strategies to ecological and environmental changes. More specifically, factors that are affecting river ecosystem health and flows as well as biological productivity were explored to ascertain any universality across the Zambezi Basin. As a prerequisite of comparative analysis, we attempted to employ comparable data sets and case studies across the river catchments concerned; however, the negative did not necessarily form a basis for non-comparison but rather strengthened cases particularly in discussion as explained in detail in the following section, which looks at the comparative research method. Based on these results and findings, recommendations are made to help increase awareness, research practices and adaptation capacity of the riparian communities to ecological changes.

Our Approach to the Comparative Research Method

In an effort to disentangle ourselves from the mess that often arises from the complications of the CRM as described earlier, we used what we considered a simplified multi-disciplinary approach. We use the definition by Atlas.eu (2015), that, 'a multidisciplinary approach involves drawing appropriately from multiple disciplines to redefine problems outside normal boundaries and reach solutions based on a new understanding of complex situations; i.e. more holistic than reductionist'. Primary qualitative and quantitative and secondary elements of study were considered. Therefore, a mixed approach, specifically the concurrent triangulation method (Hantrais 1995) was used where multiple methods were applied at different stages of the research (Bernard 2006). The approach used pre-existing secondary data, questionnaires, field studies designed to capture river health in the basin, local indigenous knowledge systems and practices (LIKSP) and historical climate trends and future modelling. The research designed methods invoked a participatory approach. The value of qualitative data also lies in understanding more than measuring differences. In addition, we considered the data available on a case by case basis, subdivided it into themes (ecology, hydrology, socio-economic settings with emphasis on gender, livelihoods, indigenous knowledge and policy dynamics) and, where possible, used historical analysis to ascertain any changes over time as well as spatial similarities and differences (across nations). This was so that we could answer the overarching question of ecological changes and the factors determining these changes spatially and temporally. As advocated for by Poteete et al. (2010) we took advantage of the expertise provided by the collaborative and trans-disciplinary nature of our research team to lean on specific methods that could be applied across the research themes which also ensured rigour across the different disciplines of study and, of course, based on available data (see CODESRIA 2015).

The strength of the present study and the approaches used, particularly the CRMs and Landscape Approach (LA) as well as Integrated Landscape Management (ILM) and ecosystem approaches, is their holistic nature—i.e., the ability to employ both qualitative and quantitative approaches, and being multi-disciplinary. Such a strategy builds on participatory approaches and capitalizes on the diverse disciplines (agriculture, ecology, fisheries, hydrology, scientists' and extension's knowledge, social, economic and political sciences) and their products and experiences in bringing out positive changes in people's livelihoods. In addition, our approach employs the Participatory Rural Appraisal (PRA) concept to share ideas and understand issues from the perspectives of the local communities. This is extremely unique and important as often the environment is not studied in tandem with its inhabitants. The primary data collection was conducted in the Manyame Basin in Zimbabwe and, where possible, comparative analyses were conducted using secondary data from similar studies in the Zambezi River Basin; that is in the Barotse Flood Plain (Zambia) and Ruhuhu and Songwe (Tanzania) and the Shire (Malawi) River catchments. Moreover, the CRM was used as a cross-cutting analytical tool not only in statistical analyses but more in discussing the findings. The specific methodologies for each respective chapter are outlined in detail therein; however, the general methods are described below.

Objectives and Research Questions

Objectives

The main objective of the study was to assess the possible changes in ecological health of selected rivers in the Zambezi River Basin and the resultant goods and services for the sustenance of rural livelihoods. The specific objectives were to:

i. Explore the different capital and livelihood base of the communities in the five river catchments under study (Chapters 3 and 5);

ii. Assess perceptions and indigenous knowledge of communities regarding ecological changes vis a vis aquatic ecosystem goods and services (Chapters 3, 4, 5, 6, 7 and 8);

iii. Evaluate the historical, current and possible future drivers (including climate changes/variability) of ecological changes and their impacts on the ecological integrity of aquatic systems and consequently on the goods and services for the sustenance of livelihoods of the basin's inhabitants (Chapters 3, 4, 5, 6, 7 and 8);

iv. Explore response mechanisms currently in place and suggest measures that should be put in place to address the effects of current and future ecological changes (Chapters 3, 4, 5, 6, 7 and 8).

Research Questions

In line with the objectives, the following research questions were posed:

i. How do the current and future drivers (if at all climate-mediated) influence changes in river flow regimes and in turn affect river ecosystems and environmental sustainability in the river basins under study?

ii. In turn, to what extent do the identified impacts compromise goods and services and human livelihoods, if at all?

iii. Are these impacts on human livelihoods by climate-induced and other factors the same across rivers/tributaries within the Zambezi Basin?

iv. What are the similarities and differences in response mechanisms aimed at addressing ecological changes across the communities along the Barotse Flood Plain (Zambia), Manyame River (Zimbabwe), Ruhuhu and Songwe Rivers (Tanzania), and Shire River (Malawi) in spite of differing socio-cultural settings?

Research Methodology
Sampling Tools, Instruments and Analyses

A structured questionnaire with open-ended questions, and focus group discussions were used to interview the communities, in order to assess their perceptions on the benefits from rivers, ecological changes, climate-induced changes, and related adaptation mechanisms. These data were analysed using the Interpretive Phenomenological Analysis (IPA), a method which offers insights into how a given person, in a given context, makes sense of a given phenomenon. Usually these phenomena relate to experiences of some personal significance such as major life events, or the development of an important relationship. The data was categorized according to: community perceptions, Indigenous Knowledge Systems (IKS) regarding ecological and climate changes and variability, impacts from these changes as well as the institutional framework and responses to each identified impact. The primary data analysis was also used to triangulate thematic data gathered from secondary sources.

A participatory historical trend analysis to understand the extent of the natural resource base that these communities depend on was performed in collaboration with the communities. This analysis was also intended to understand the extent to which the natural resource base has changed over time. This exercise was preceded by community resource mapping, which provided an understanding of the general natural resource base at community level and also issues of access and use by gender and other socio-economic groups in the study areas. Transect walks and transect belts complemented the already highlighted activities and helped witness first-hand the extent of the resource base as well as to gauge the changes that were noted in a participatory way. The walks were also used to validate data for access and use issues that were highlighted. Secondary desk studies through literature and partner studies were used to identify data gaps and validate the data gathered.

In addition, plankton (diatoms), macroinvertebrates and fish inventories were undertaken in and around the rivers as part of the natural resources baseline data collection. Diatoms, macroinvertebrate and fish genera and species were identified and enumerated in selected sites around the Zambezi basin, i.e., in and around the sampling area where the sampled communities live. More so, already existing (secondary) data was sought from research organizations and non-governmental organizations (NGOs) that have worked on flora and fauna in the study areas. This enabled the determination of spatial and temporal changes in biota and also allowed comparisons with historical data. In addition, the use of biological monitoring in assessing the health of these river systems was assessed with the intention of seeking validation and, to some extent, the calibration of local rivers to external river bio-monitoring or assessment tools (see Chapter 4).

Furthermore, a water resources assessment of the study area was conducted and included analysis of existing hydrological and meteorological databases from water regulatory authorities and meteorological offices in the respective countries. The data were used to calculate means, anomalies, and variability of water resources and climate. In addition, both climate and hydrological modelling were employed to understand climatic and hydrological processes in the basin (see Chapters 5 and 6). Responses from interviews coupled with field measurements of biophysical indicators for specific water sources were used to determine river water quantity and quality and availability including on-site information (water sources, watering points, evidence of runoff, flooding and river flood pulsing) and wider off-site or ecosystem effects of land use / management practices, including the effectiveness of water management techniques by the community and how this interfaces with other resources (see Chapters 6 and 7).

Some climate analyses were attempted where seasonal and annual trends in air temperatures and rainfall were determined using data from respective countries' Meteorological Services; and parallels were drawn with community perceptions regarding climate change and variability. Future climate scenarios investigated were conducted using climate model simulations. The performance of climate models was tested (see Chapter 5). In addition, possible impacts of climate (rainfall and temperature) on hydrological factors (water supply, runoff, river water levels and water flows) in the Zambezi Valley and, subsequently, the impact on livelihoods, agricultural and ecosystem production were examined. Trend and extreme value analysis were performed on hydrological datasets to verify the occurrence of climate change and accompanying impacts on the maintenance of the hydrological regime. A determination of whether the hydrological regime is affected by floods and to what extent this impacts environmental flows and the accompanying impacts on basin ecology was made through statistical analysis of historical and current rainfall levels, river water levels and flow characteristics (see Chapter 6).

Trend Analysis of Water Discharge in the Zambezi River Basin

Trend analysis is normally undertaken in order to discern the direction of hydrological or climatological time series data. Mann-Kendall is a statistical test usually used for trend analysis in hydrological and climatology time series. This method was chosen in accordance with Millard (2013) as it can be used with some missing data in the flow records, which is commonplace in Sub-Saharan Africa. This test statistic is used to test the null hypothesis, H_o if $|Z_s|$ is greater than $Z_{a/2}$, with significance level (e.g., 5 per cent with $Z_{0.025} = 1.96$) then the null hypothesis is invalid, implying that trend is significant. The Kruskal-Wallis non-parametric test was used to determine statistically significant decadal changes in temperature and precipitation in the five selected Zambezi River sub-catchments and a Mann-Kendall test was used to determine seasonal and annual climate (precipitation and temperature) trends (see Chapter 5). Historical trends of the discharge data obtained from different sources and covering Lake Malawi/ Nyasa catchments in Tanzania, Shire River Basin in Malawi, Upper Zambezi Sub-Basin and Manyame catchment in Zimbabwe were determined using the Mann-Kendall test statistic (see Chapter 6).

Conclusion

This chapter describes the methodology, and the conceptual and theoretical framework of the study. We introduce the Zambezi River in brief by discussing some environmental and anthropogenic factors, namely ecological and hydrological characteristics of the river, climate and in particular water levels and flows, which are in turn affected by human pressures such as increasing population and therefore demand on water use and supply. In addition, a description of the Barotse Flood Plain, Manyame River, Ruhuhu River, Shire River and Songwe River—which are the lotic systems under study is presented, and includes their location, physiography, climate and land use.

The Comparative Research Method (CRM) is discussed as the common analytical framework used across the study which can be used in cross-cultural and trans-boundary research. The limitations of the CRM reflected that despite the challenges (mainly methodological) that can be encountered in engaging with it, there are still several advantages of using it which embody the holistic approach to research. CRMs are useful in depicting particular issues; dealing with the analysis of cross-cultural data as well as reviewing the main issues in the methodology. In addition, they are also helpful in pushing cross-cultural researchers to place more emphasis on methods of data analysis to improve the effectiveness of studies while dispelling the myth that methodological and statistical sophistication is an obstacle or a distraction in the research enterprise. Our mixed approach to comparative research includes the concepts of the Landscape, Integrated Landscape and Ecosystems approaches, and the use of both qualitative and quantitative data as well as empirical and anecdotal evidence. In this effort we stretched the CRM with the hope that, some good analyses and results would still be achieved. All these will be discussed exhaustively in the ensuing chapters.

References

Alasuutari, P., 1995, Researching Culture. Qualitative Methods and Cultural Studies, London: Sage Publications/California, USA: Thousand Oaks.

Ashton, P. J., Love, D., Mahachi, H. & Dirks, P. H. G. M., 2001, 'An overview of the impact of mining and mineral processing operations on water resources and water quality in the Zambezi, Limpopo and Olifants Catchments in Southern Africa', Contract Report to the Mining, Minerals and Sustainable Development (Southern Africa) Project, by Report No. ENV-P-C 2001-042. Harare, Zimbabwe: CSIR-Environmentek, Pretoria, South Africa and Geology Department, University of Zimbabwe, xvi + 336 pp.

Atlas, eu, 2015, 'Multi-disciplinary approach (of study, design)', Article accessed from http://atlas.uniscape.eu/glossarioToto.php?idgl=30&nomegl=Multidisciplinary%20 Approach%20(of%20study,%20of%20research)&lettera= on 01 February 2016.

Beilfuss, R. & Brown, C., 2010, 'Assessing environmental flow requirements and trade-offs for the Lower Zambezi River and Delta, Mozambique', *International Journal of River Basin Management*, 8, 127-138.

Bere, T. & Mangadze, T., 2014, 'Diatom communities in streams draining urban areas: community structure in relation to environmental variables', *Journal of Tropical Ecology*, 55 (2), 271-281.

Berman, E. M., 2002, Essential Statistics for Public Managers and Policy Analysts, Washington, DC.

Bernard, H. R. 2006, Research methods in anthropology, Oxford: Alta Mira Press, 298 pp.

Bootsma, H. A. & Jórgensen, S. E., 2004, 'Lake Malawi/Nyasa', In M. Nakamura, (ed.), Managing Lake Basins: Practical approaches for sustainable use, http://www.world-lakes.org/ uploads/ELLB%20Malawi-NyasaDraftFinal.14Nov2004.pdf Downloaded on 28 August 2015, 36 pp.

Bootsma, H. A., 2006, 'Lake Malawi/Nyasa. Experience and lessons learnt. Lakes and Reservoirs: *Research and Management*, 11, 271-286.

Brammer, H., 1973, 'Soil profile descriptions and analytical data and an account of soil genesis and classification. Chilanga, Zambia: Department of Agriculture', 320.

Branchu, P., Bergonzini, L., Ambrosi, J.-P., Cardinal, D., Delalande, M., Pons-Branchu, E. & Benedetti, M., 2010, 'Hydrochemistry (major and trace elements) of Lake Malawi (Nyasa), Tanzanian Northern Basin: local versus global considerations', *Hydrology and Earth System Sciences Discussions*, 7, 4371-4409.

Brislin, R. W., 1970, 'Back translation for cross cultural research', *Journal of Cross-cultural Psychology*, 1, 185-216.

CGIAR CRP on AAS, 2012, 'Barotse Floodplain, Zambia', Scoping Report. https://
    en.wikipedia. org/wiki/Special: Book Sources/978-0-566-05196-8 Downloaded on
    27 August 2015.

Chafota, J., 2012, 'Integrated natural resources management in a dynamic transbound-
    ary watershed context: the Songwe River catchment experience', WWF, CDE and
    ASDC,121 pp.

Chale, F. M. M., 2010, 'Preliminary studies on the ecology of Mbasa (Opsaridium mi-
    crolepis (Gunther) in Lake Nyasa around the Ruhuhu River', Journal of Ecology and
    the Natural Environment, 3, 58-62.

Chale, F. M. M., 2011, 'Preliminary study of the ecology of mbasa (Obsaridium micro-
    lepsis (Gunther) in Lake Nyasa around the Ruhuhu River', *Journal of Ecology and the
    Natural Environment*, 3 (2), 58-62.

Chiuta, T. M. & Johnson, S., 2010, 'Songwe River Transboundary Catchment Man-
    agement Project (Tanzania-Malawi)', Final Evaluation Report, Swiss Development
    Cooperation and WWF, 50 pp.

CODESRIA, 2015, 'Comparative research networks 2015 competition', Retrieved Au-
    gust 19, 2015, from CODESRIA: http://www.codesria.org/spip.php?article2243.

Crul, R. C. M., 1997, 'Limnology and hydrology of Lakes Tanganyika and Malawi',
    Studies and reports in hydrology 54, Paris: UNESCO, 105 pp.

Darwall, W., Smith, K., Allen, D., Holland, R., Harrison, I., Brooks, E., Pacheco Chaves,
    B., Springer, M., Sermeño Chicas, J.M., Sermeño Chicas, J.M. & Pérez, D., 2008,
    'The diversity of life in African freshwaters: underwater, under threat. An analysis
    of the status and distribution of freshwater species throughout mainland Africa'
    (No. 333.95096 D618div). San Salvador (El Salvador): Universidad de El Salvador.

Duponchelle, F., Ribbink, A.J., Msukwa, A., Mafuku, J. & Mandere, D., 2000. 'The
    potential influence of fluvial sediments on rock-dwelling fish communities', in F. Du-
    ponchelle, & A.J. Ribbink, Eds, Fish ecology report. Lake Malawi/ Nyasa/ Niassa
    Biodiversity Conservation Project, SADC/GEF, pp.111-132.

Flint, L., 2009, 'Climate change, vulnerability and the potential for adaptation:
    case-study-the Upper Zambezi Valley region of Western Zambia', in P.S. Ranade, ed,
    Climate Change: Impact and mitigation. Hyderabad/ Zambia: Icfai University Press.

Fisheries Act, (2011). No 22 of 2011. Lusaka, Zambia: Government of Zambia, pp. 379-
    418. Hantrais, L., 1995, 'Comparative Research Methods', *Social Research Updates*,
    13, 1-3.

Hranova, R., Gumbo, B., Klein, J. & Van Der Zaag, P., 2002, 'Aspects of the water
    resources management practice with emphasis on nutrients control in the Chivero
    Basin, Zimbabwe', *Physics and Chemistry of the Earth*, Parts A/B/C, 27, 875-885.

Hirsch, R. M., Helsel, D. R., Cohm, T. A. & Gilroy, E. J., 1993, Statistical analysis of
    hydrological data, United Kingdom: McGraw-Hill Incorporated.

IUCN, 2003, 'Barotse Floodplain, Zambia: Local economic dependence on wetland re-
    sources', Case Studies in Wetland Valuation #2. Gland, Switzerland: IUCN.

IUCN, 2006, The Red list of Threatened species. Gland, Switzerland: IUCN.

IUCN, 2012, Livelihoods and Landscapes Strategy: Results and Reflections, Gland,
    Switzerland.

IUCN, 2015, 'The redlist of threatened species', Version 2015.2. www.iucnredlist.org. Downloaded on 13 August 2015.

Jessen, O. Z. & Von Christierson, 2015, 'Shire basin institutional planning and DSS. Groundwater Assessment Report', Niras Technical Report, July 2015, pp. 1-95.

Kambole, M. S., 2003, 'Managing the water quality of the Kafue River', *Physics and Chemistry of the Earth*, Parts A/B/C, 28, 1105-1109.

Kendall, M. G. & Stuart, A., 1968, The advanced theory of statistics, Design and analysis, and time series, London: Charles Griffin and Co. Ltd.

Kidd, C. H. R., 1983, 'A Water Resources Evaluation of Lake Malawi and the Shire river', in Project, U., ed., Geneva: World Meteorological Organisation, pp.132.

Kingdon, M. J., Bootsma, H. A., Mwita, J. & Mwichande, B., 1999, 'River discharge and water quality', in H. A. Bootsma & R. E. Hecky, eds, Water Quality Report, Lake Malawi/ Nyasa Biodiversity Conservation Project. SADC/GEF, pp. 29-69.

Kirchhoff, T., Trepl, L. & Vicenzotti, V., 2013, 'What is landscape ecology? An analysis and evaluation of six different conceptions', *Landscape Research*, 38(1), 33-51. DOI:1 0.1080/01426397.2011.640751.

Kreuser, T., 1991, 'Facies evolution and cyclicity of alluvial coal deposits in the Lower Permian of East Africa (Tanzania)', *Geologische Rundschau*, 80, 19-48.

Lake Nyasa Basin Water Board (LNBWB), 2014, Basin annual hydrological report. Tukuyu, Tanzania: Ministry of Water-The United Republic of Tanzania.

Lincoln, Y. S. & GUBA, E. G., 1985, Naturalistic Inquiry, Beverly Hills, CA: Sage Publications.

Lotz-Sisitka, H. & Urquhart, P., 2014, 'Strengthening university contributions to climate compatible development in Southern Africa: Zimbabwe country report' in P. Kotecha, ed., SARUA Climate change counts mapping study, South Africa: Witswatersrand.

Magadza, C. H. D., 2003, Lake Chivero: A management case study, Lakes and Reservoirs, *Research and Management*, 8, 69-81.

Makore, G. & Zano, V., 2012, 'Mining within Zimbabwe's Great Dyke: extent, impacts and opportunities. Harare, Zimbabwe: Zimbabwe Environmental Law Association (ZELA)'. http://hrbcountryguide.org/wp-content/uploads/2013/10/Mining-within-Zimbabwes-Great-Dyke.pdf (Viewed on 24 July 2016).

Millard, S. P., 2013, Environmental Statistics: An R Package for Environmental Statistics, New York, USA: Springer-Verlag.

Ministry of Environment and Natural Resources Management, 2013, 'Zimbabwe National Climate Change Response Strategy (First Draft)'. Harare, Zimbabwe. http://www.ies. ac.zw/downloads/draft%20strategy.pdf. Downloaded on 6 July 2015.

Ministry of Finance, 2013, 'Investment projects for the Barotse and Kafue sub-basins under the Strategic Programme for Climate Resilience in Zambia Environmental and Social management framework', Final Draft report, NIRAS, Zambia.

Moore, A.E., Broderick, T. & Plowes, D. 2009, Landscape evolution in Zimbabwe from the Permian to present, with implications for kimberlite prospecting, *South African Journal of Geology*, 112(1), 65-88.

Morse, J. M., 2003,'Biasphobia', *Qualitative Health Research*, 13, 891-892.

MOW, 2013a, 'Preparation of an integrated water resources management and development plan for the Lake Nyasa Basin', Climate change report. Prepared by SMEC, pp. 49.

MOW, 2013b, 'Preparation of an integrated water resources management and develop-ment plan for the Lake Nyasa Basin: river health report', Prepared by SMEC, pp.29.

MOW, 2013c, 'Preparation of an integrated water resources management and development plan for the Lake Nyasa Basin', Water Quality Report, Prepared by SMEC, pp. 32.

Moyo, N. A. G., 1997, Lake Chivero, a polluted lake, International Union for the Conservation of Nature and Natural Resources/University of Zimbabwe Publications, Harare.

Mudziwapasi, R., 2015, Re: Personal Communication, 'Biri Dam Fish Farming Cooperative Group Member, Chinhoyi, Zimbabwe', Type to M. R. Ndebele-Murisa. Mugandani, R., Wuta, M., Makarau. A. & Chipindu, B., 2012, 'Re-classification of agro-ecological regions of Zimbabwe in conformity with climate variability and change', *African Journal of Crop Science*, 20, 361-369.

Nanthambwe, S., 2013, 'Policy sector review for incorporating sustainable land management in the shire river basin and development of an institutional framework for sustainable land management', report submitted to the UNDP and the Government of Malawi, Lilongwe, Malawi.

NBI, 2007, 'Strategic/sectoral social and environmental assessment of power development options in the Nile Equatorial Lakes Region' Final Report Vol. 1 Main Report, SNC-Lavalin International.

Ndebele, M. R. & Magadza, C. H. D., 2006, 'The occurrence of Mycrocystin-LR in Lake Chivero, Zimbabwe', Lakes and Reservoirs: *Research and Management*, 11, 57-62.

Nhamo, L. & Chilonda, P., 2012, 'Climate change risk and vulnerability mapping and profiling at local level using the Household Economy Approach (HEA)', *Earth Science and Climatic Change*, 3,3-7. http://dx.doi.org/10.4172/2157-7617.1000123.

Nhapi, I., Holko, Z., Siebel, M. A. & Gijzen, H. J., 2002, 'Assessment of the major water and nutrient flows in the Chivero catchment area, Zimbabwe', *Physics and Chemistry of the Earth*, Parts A/B/C, 27, 783-792.

Nicholson, S. E., Klotter, D. & Chavula, G., 2014, 'A detailed rainfall climatology for Malawi, southern Africa', *International Journal of Climatology*, 34(2), 315-325.

Nindi, S. J. 2007. Changing livelihoods and the environment along Lake Nyasa, Tanzania, *African Study Monographs*, 36, 71-93.

Odada, E. O., Olago, D. O., Bugenyi, F., Kulindwa, K, Karimumuryango, J., West, K., Ntiba, M., Wandiga, S., Aloo-Obudho, P. &Achola, P., 2003, 'Environmental assessment of the East African Rift Valley lakes', *Aquatic Sciences Research Across Boundaries* 65, 254-271.

Pike, K., 1967, Language in relation to a united theory of the structure of human behavior, Mouton: The Hague.

Poteete, A. R., Janssen, M. A. & Ostrom, E., 2010, Working Together: Collective Action, the Commons, and Multiple Methods in Practice, New Jersey, Princeton: Princeton University Press.

Proctor, J. & Cole, M. 1991. The ecology of ultramaphic rocks in Zimbabwe. The Ecology of Areas with Serpentinised Rocks. 313-331p.

Sayer, J. A., Sunderland, T. C. H., Ghazoul, J., Pfund, J. L., Sheil, D., Meijard, E., Venter, M., Boedhihartono, A. K., Day, M., García, C., Van Oosten, C. & Buck, L. E., 2013,

'Ten principles for a landscape approach to reconciling agriculture, conservation, and other competing land uses', *Proceedings of the National Academy of Sciences* (PNAS), 110, 8349-8356.

Shire River Basin Atlas, 2012, Malawi National Spatial Data Centre, 2012, pp. 1-66.

SMEC, 2013, Preparation of an integrated water resources management and development plan for the Lake Nyasa Basin: River health report. URT, Ministry of Water, 29 pp.

Thornton, J. A. & Nduku, W. K., 1982, Lake McIlwaine: The eutrophication and recovery of a tropical lake, Dr. W. Junk, The Hague.

Timberlake, L., 1998, 'Biodiversity of the Zambezi Basin Wetlands: Review and preliminary Assessment of available information', in: IUCN, (ed., Consultancy Report for IUCN, The World Conservation Union, Regional Office for Southern Africa. Harare, Zimbabwe.

Turner, G., 1995, 'Management, conservation and species changes of exploited fish stocks in Lake Malawi' in T. J. Pitcher, & Hart, P. J. B. (eds.) The Impact of Species Changes in African Lakes. Chapman & Hall Fish and Fisheries Series 18, Springer Netherlands, 365-395.

URT, (2013), 2012, 'Population and housing census population distribution by administrative areas', National Bureau of Statistics. Dar es Salaam and Zanzibar: Ministry of Finance, 1-264 pp.

Utete, B., Nhiwatiwa, T., Barson, M. & Mabika, N., 2013, 'Metal correlations in sediment and in water from the Gwebi catchment, Zimbabwe', *International Journal of Water Science*, 2, 1-8.

Vincent, V. & Thomas, R. G., 1960, An Agricultural Survey of Southern Rhodesia: Part I: Agro-ecological survey. Salisbury, Rhodesia: Government Printers. Water Department/UNDP, 1986, National Water Resources Master Plan, in G.K.

Kululanga & G.M.S. Chavula, 1993, 'National Environmental Action Plan: A Report on Water Resources', Submitted to Ministry of Research and Environmental Affairs,. Blantyre, Malawi: Malawi Polytechnic.

Weiner, I. B., Freedheim, D. K. & Graham, J. R., 2003, Handbook of Psychology, New Jersey, USA: John Wiley and Sons.

World Bank, 2002, 'Project ID for Lake Malawi ecosystem management Project. Project ID MWPE P066196', Washington DC: The World Bank.

Wu, J., 2006, 'Cross-disciplinarity, landscape ecology, and sustainability science', *Landscape Ecology*, 21, 1-4.

Yu, Y. S. & Zuo, S., 1993, 'Relating trends in water quality of principal components to trends of water quality constituents', *Water Resources Bulletin*, 29, 797-806.

Zawa, 2006, 'Zambezi Floodplains, Information Sheet on Ramsar Wetlands', (RIS)-2006-2008 Version.

Zimbabwe National Statistics Agency, 2012, Zimbabwe Population Census 2012, National Report, Harare, Zimbabwe.

3

Trends in Ecological Changes: Implications for East and Southern Africa

Ismael Aaron Kimirei, Chipo Plaxedes Mubaya,
Mzime Ndebele-Murisa, Lulu Kaaya, Tinotenda Mangadze,
Tongayi Mwedzi and Judith Natsai Theodora Kushata

Introduction

Over the past 50 years, changes in the earth's ecosystems have been noted, particularly in developing countries and mostly due to the high demand for resources caused by the ever-growing human population as well as by climate change and variability (McCarty 2001; Millennium Ecosystem Assessment 2005; UN-DESA 2011 IPCC 2014). The demand for fish, freshwater supply for domestic and industrial use and hydropower production, and food to sustain the human population is increasing by the day (Lehane 2013). In fact, it has been estimated that by 2025, 1.8 billion people will be living in regions with absolute water scarcity; almost half of the worlds' population (between 75 million and 250 million people in Africa) will be living in places of intense water stress, displacing between 24 million and 700 million +in arid and semi-arid places (UNEP 1999; UNDP 2006; UN-Water and FAO 2007; UN-Water 2013). As a result of these increasing demands, humans are modifying the natural environment in order to sustain their needs and this subsequently can hamper the ability of ecosystems to provide benefits to communities as well as self-recuperate from damage. This is disconcerting given that it is these same ecosystems and their biodiversity whose contribution to human well-beings is unmatched (Mutasa and Ndebele-Murisa 2015). Aquatic ecosystems in particular provide priceless goods and services without which survival on earth would be seriously limited.

In most cases, we exploit the goods and services provided by the ecosystems only for short-term gains while changing them almost forever, causing effects that are either impossible or very hard to reverse because of accumulated and unnoticed long-term costs (Millennium Ecosystem Assessment 2005).

Many tropical aquatic ecosystems are ecologically stressed mainly from unsustainable land uses, poor agricultural practices, and deforestation of the catchments, pollution from domestic sources and mines, overfishing, and climate changes. At the centre of all these ecosystem stressors is the high human population growth rate, with the global human population expected to hit the 10.6 billion mark by 2050 (UN-DESA 2011); the human population in Sub-Saharan Africa will be about 2.2 billion by the same year (UN-DESA 2011). This will further intensify the pressure on aquatic ecosystems worldwide (Millennium Ecosystem Assessment 2005), and in the region. Moon (2011), the then United Nations Secretary General, stated that competition between communities and countries for scarce resources, especially water is increasing, as well as environmental refugees. This is reshaping the human geography of the planet, a trend that will only increase as deserts advance, forests continue to be felled and sea levels continue to rise. Unsustainable human activities in and around the world's freshwater ecosystems, i.e. rivers, streams, lakes and wetlands alike and the increasing demand for natural resources are exacerbating the already high strain on these ecosystems. In the east and southern African region, agricultural lands are increasingly being located around water bodies as these regions try to attain food security (Tumbare 2004).

Most aquatic ecosystems in east and southern Africa are negatively impacted through the deterioration of water quality, mainly caused by sedimentation (O'Reilly 1998; Thrush et al. 2003; Tumbare 2004; Nindi 2007) and excessive nutrients input, loss of critical habitats for fish and other organisms and the overall decline in ecological conditions (Millennium Ecosystem Assessment 2005; Chale 2010; Teresa and Casatti 2012). It is also important to note that the survival of many aquatic organisms such as fish and invertebrates depends on the health of not only the habitats they live in but also their immediate riparian zones (Holmes 2010). Degradation of the physical structure of rivers, for example, can have far-reaching consequences on habitat diversity and hence fish and invertebrates. The sanctity or keeping aquatic ecosystems healthy is vital for continued provisioning of ecosystem goods and services that communities and different ecological facets accrue. Sedimentation from agricultural land expansion and deforestation (Tumbare 2004), nutrients loading from untreated sewage, wastewater from industries, atmospheric deposition and agricultural fertilizers (Bootsma et al. 1999; Ashton et al. 2001; Tumbare 2004) all remain a challenge to aquatic ecosystems in east and southern Africa. In addition, unsustainable fishing activities, overexploitation of aquatic resources, and the introduction of exotic species are changing the ecological settings and the balance of many of the ecosystems in the region (Ogutu-Ohwayo 1990; Kudhongania et al. 1992; Witte et al. 1992; Turner 1995; Millennium Ecosystem Assessment 2005). Introduction of exotic species, for example, has devastated the ecology of Lakes Kariba, Kivu and Victoria by changing species composition, dominance, diversity and probably some extinction of indigenous species (Ogutu-Ohwayo 1990; Kudhongania et al. 1992; Witte et al. 1992; Witte et al. 2009). This chapter explores trends in ecological changes in east and southern Africa with an emphasis on the Zambezi River Basin and highlights a case study of these changes as perceived by local communities within the Manyame catchment in Zimbabwe.

States, Trends and Causes of Ecological Changes

Ecological changes entail changes in biotic and abiotic factors. In this case, changes in the environment as well as biotic interactions within rivers and lakes ultimately affect the biophysical environment and interactions between humans and the environment. These changes can be shifts in species composition, distribution, and population dynamics or a complete alteration of the species flock; changes in water quality, physical habitat change and human ecological interactions. Ecological changes can be monitored or assessed using various indicators, such as fish species change (species, size, diversity and production), riparian vegetation cover and distribution, shifts in vegetation species diversity, macro-and micro-invertebrates diversity and distribution, metal pollution, eutrophication and nutrients enrichment in aquatic systems, phytoplankton and zooplankton dynamics, and sediment accretion rates (see Chapter 4 of this volume).

The ecological status of many freshwater ecosystems in the east and southern Africa region, and the Zambezi River Basin in particular, has changed significantly and in many ways since the beginning of the twentieth century. For example, the undercutting of the Zambezi channel and the reduction of the water table in the Zambezi River flood plain have been noticed (Benedict et al. 2007; Ndhlovu 2013). Fish species composition and diversity changes have been registered in Lakes Nyasa/Malawi, Tanganyika, Victoria, Kariba and several rivers within the Zambezi River Basin (Turner 1995; Kolding et al. 2003; Kimirei and Mgaya 2007; Kimirei et al. 2008; Kayanda et al. 2009; Witte et al. 2009; Chale 2010; IUCN 2010; Marshall 2011; Ndebele-Murisa et al. 2011a; Ndebele-Murisa et al. 2011b). Also, heavy metal pollution, mainly from mining activities and manufacturing industries, land use and cover changes (Odada et al. 2003; Millennium Ecosystem Assessment 2005; Nelson et al. 2006) as well as phytoplankton and zooplankton diversity and composition changes have all been reported in the region's waters (Kurki et al. 1999; Descy et al. 2005; Descy et al. 2010; Ndebele-Murisa et al. 2010; Ndebele-Murisa et al. 2012). While the anthropogenic modifications of the aquatic biophysical environment are threatening ecosystem functioning, natural as well as global environmental changes are threatening the future of many of the Zambezi River Basin's inhabitants, inclusive of human beings (McCarty 2001; Rosenblatt 2005). For this reason, it becomes important to not only focus on mitigation of environmental changes but also the regulation of human activities and to ensure that these activities are kept in check through research-informed policy making processes and implementation.

The Millennium Ecosystem Assessment (2005) report indicates that human-induced changes, such as land use and cover change, habitat change and overexploitation of natural resources (fish, timber, minerals), water abstraction and climate change are at the centre of ecosystem changes. In many rivers and lake catchments within the east and southern African region, erosion and sedimentation are major problems. The extractive industry, for example, is growing at an alarming pace in the region (Ashton et al. 2001; Shoko and Love 2005; Kamete 2007; Mudimbu et al. 2012), while irrigated farmlands are expanding. Exploration for gas and oil, various other minerals and rare metals, and extraction of the same are adding an extra dimension on ecological problems, and actually are expected to devastate provision of ecosystem goods and services and cause transboundary conflicts (Ashton et al. 2001; Shoko 2002). Untreated effluents from mines within the Zambezi River Basin are sometimes released into the aquatic ecosystems, thereby polluting not only the immediate recipient systems but more so the receiving waters, which are normally lakes and reservoirs. For example, mining activities in the upper catchments of Mngaka and Njombe Rivers in Tanzania pose threats of acid mine drainage into the Ruhuhu River (LNBWB 2014). Artisanal and large-scale mining in the Ruhuhu catchment, which includes alluvial gold, coal and iron mining, threatens to impair the water quality of Lake Nyasa/Malawi, particularly by lowering pH levels which may have tremendous impacts on aquatic life. In Botswana, alluvial gold panning is booming and consequently, ecological disasters are certainly looming.

In and around the Okavango River Basin (Botswana), a major tributary of the Zambezi for example, mining activities threaten the groundwater quantity and quality. Groundwater contamination has already been reported in Botswana, where 340 (17 per cent) out of 2,000 boreholes studied showed signs of increasing total dissolved solids and nitrate pollution, especially in densely populated areas of south-eastern Botswana (Kgathi 1999). Mining in the Okavango River Basin has even resulted in habitat change and loss of biodiversity (plant life). Unsightly dumps of ore have caused solid wastes and pollution of groundwater, while sand mining in Angola and Namibia has been linked to increased salinization in the delta. Furthermore, it is possible that the Boteti River (Botswana) stopped flowing as a result of mining (Mmopelwa et al. 2011). Heavy metal pollution in the entire Zambezi River Basin is inevitable given the current rush for gold in the basin. For example, mercury (Hg) concentrations of up to 0.21 μgl^{-1}, which is beyond the WHO standards for drinking water, have been reported in various river catchments within the Zambezi River Basin (Shoko 2002; Chifamba 2007; Nakayama et al. 2010; Nhiwatiwa et al. 2011; Mabika et al. 2015). The gravity of the associated ecological consequences cannot be over-emphasized.

Introduction of species has been both an ecological blessing and a curse to aquatic systems and communities in east and southern Africa. For example, Lakes Kariba and Victoria's ecological state and fisheries biodiversity may take some decades to equilibrate and or may never return to the state prior to the Nile perch and other species' introductions (Ogutu-Ohwayo 1990; Kudhongania et al. 1992; Witte et al. 2009); especially with increasing eutrophication and overexploitation (Hecky et al. 1994; Odada et al. 2003; Ogutu-Ohwayo and Balirwa 2006). In addition, Lake Victoria became eutrophic due to both excessive nutrients and sediment inputs from poorly managed land use coupled with changes in species composition as a result of displacement by the introduction of Nile perch and other species (Hecky et al. 1994; Odada et al. 2003). The fishery of Lake Victoria has changed from a multispecies to one dominated by the two introduced species—*Lates niloticus* and *Oreochromis niloticus* and one native cyprinid—*Rastrineobola argentea* (Kayanda et al. 2009). Introduction and invasions in Lake Kariba, on the other hand, were perceived as a blessing due to increased fish catches as a result of the introduction of two fish species— *Limnothrissa miodon* (Kapenta—the Tanganyika sardine) and *O. niloticus* (Nile bream) (Chifamba 2000; Kolding et al. 2003; Chifamba and Videler 2014). However, recent, negative changes in fish catches and composition in Kariba are raising concerns (Zengeya and Marshall 2010; Ndebele-Murisa et al. 2011a). Magadza (2006) and Tumbare (2008) identify exotic and invasive species, namely *L. miodon* and the noxious weed *Eichhornia crassipes* as major threats to Lake Kariba's ecosystem. More recently, the Australian crayfish (*Cherax quadricarinatus*) seems to be causing some considerable impacts on fish ecology as a top predator (Marufu et al. 2014) although more extensive research is required on this fairly new exotic species.

A similar case of declines in fish catches is evident in Lakes Tanganyika and Nyasa/Malawi (Turner 1995; O'Reilly et al. 2003; Sarvala et al. 2006; Kimirei et al. 2008); all of which are linked to both climatic changes and overexploitation (O'Reilly et al. 2003; Verburg et al. 2003; Vollmer et al. 2005; Sarvala et al. 2006; Plisnier et al. 2009). The pelagic fisheries of Lake Tanganyika have changed from a six-species one to a three-species fishery which seems to be moving to a single-species fishery (Coulter 1970; Kimirei et al. 2008; Plisnier et al. 2009); while the large and economically important fish species are declining in Lake Nyasa/Malawi (Turner 1995; Banda et al. 2005; Banda 2009). The drivers, pressures, impacts, states and responses of ecological systems discussed herein are summarized in Figure 3.1.

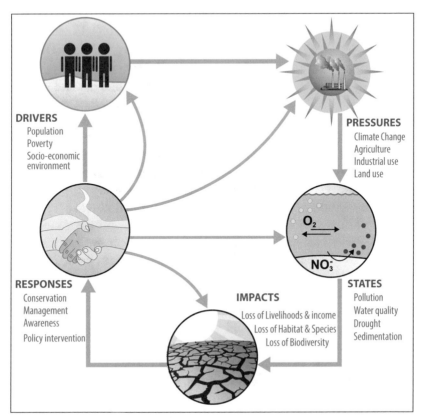

Figure 3.1: Summarized schematic presentation of drivers, pressures, state of, impacts to and responses (DPSIR) of aquatic ecosystems in the east and southern African region

The volume of anoxic waters in almost all lakes within the east and southern African region is swelling (Hecky et al. 1994; Odada et al. 2003; O'Reilly et al. 2003; Verburg et al. 2003; Msomphora 2005; Vollmer et al. 2005; van Bocxlaer et al. 2012; Ndebele-Murisa et al. 2014). Thermocline stability is increasing, mixed layers are decreasing, both surface and deep waters are warming, and transparency is increasing; all of which bear possible negative impacts on productivity in these ecosystems (Hecky et al. 1994; Hulme et al. 2001; O'Reilly et al. 2003; Verburg et al. 2003; Vollmer et al. 2005; Ndebele-Murisa et al. 2013a; Loiselle et al. 2014). In addition, lake levels have changed and/or fluctuated in Lakes Tanganyika, Nyasa/Malawi and Kariba due to climate (rainfall dynamics and drought events) (Crul 1997; Odada et al. 2003; O'Reilly et al. 2003; Vollmer et al. 2005; Cohen et al. 2006) and increasing sedimentation (Chifamba 2000; Alin et al. 2002 Nindi 2007; Chiuta and Johnson 2010, Kunz et al. 2011); which has threatened to close off some basins (Kidd et al. 1999) (Table 3.1).

Changes in land-use practices have caused many rivers in east and southern Africa to shift from perennial to seasonal (Nindi 2007). The sizes and water volumes of these rivers have significantly shrunk, and their flow directions have changed greatly since the late 1980s (Nindi 2007). For instance, the Songwe River has migrated across the Tanzania-Malawi border for years (Nindi 2007; Chafota 2012), and the Boteti River in Botswana has completely stopped flowing (Mmopelwa et al. 2011). In the Lake Nyasa/Malawi basin, catchment degradation due to farmland expansion is causing a recurrent flooding phenomenon (Kidd et al. 1999; Hecky et al. 2003; Nindi 2007). Table 3.1 summarizes some of the ecological changes that have occurred in some major lakes (Kariba, Malawi, Tanganyika and Victoria) in the east and southern African region. The table highlights these changes as mostly ecological and several authors as indicated in the table report these phenomena across the region. Although the Tanzanian part of the Lake Nyasa/Malawi catchment is still intact (MoW 2013b), a decrease in forest cover from 64 to 51 per cent between 1967 and 1990 has been reported (Calder et al. 1995). While an increase in sediment and nutrients inputs into Lake Nyasa/Malawi from deforested rivers and catchment has already been shown (Hecky et al. 2003), many more ecological and hydrological changes as a result of continued catchment degradation are anticipated. Unstable stream and river flows, changes in trophic relations and carbon flows, and decline in fish catches and breeding behaviour, all of which have negative impacts to community livelihoods, could manifest as catchment/forest degradation continues around this lake (Tweddle 1983; Turner 1995; Ramlal et al. 2003; Nindi 2007; Chale 2010; MoW 2013c; Carvalho et al. 2015). Perhaps one of the most notable and classical examples of drying up in the region comes from Lake Chilwa in Malawi which has dried up about nine times in the past century (Nyaya et al. 2011). This drying is mainly related to reduction in the annual flows which is intricately linked to climate change-mediated changes in rainfall patterns over the past 50 years; and poor land use practices in the lake catchment (Njaya et al. 2011). An exposition of climate, rainfall patterns for the four river catchments under study in this book is presented in Chapter 5, while discussions on the hydrology of some river sub-catchments within the Zambezi (rainfall, water levels and flows) are presented in Chapter 6 of this volume.

Climate Change Impacts on Ecosystems, Livelihoods and Agriculture

Ecosystems in the east and southern African region are under threat from a number of factors as expressed earlier in this chapter. These factors are likely to affect most of Africa's natural resources with potential adverse impacts on terrestrial and aquatic ecosystems (Leemans and Eickhout 2004; Boko et al. 2007). Boko et al. (2007) report that climate change impacts—such as rising temperatures and other stresses—have led to shifts of ecological zones, loss of biodiversity, and caused an overall reduction in ecological productivity. They also allude that climate change has caused species migration as a result of habitat reduction, fragmentation and loss in Africa. For plants and crops, the effect of increasing aridity and rising temperatures results in a variety of effects such as desiccation, wilting, reduction of soil nutrients and therefore reduced plant growth and productivity due to loss of soil nutrients through leaching, increased transpiration rates from most plant canopies whose rates may be reduced by the closing of stomata due to increased concentrations of carbon dioxide in the atmosphere (Kirschbaum 2000). For animals, negative impacts of the same climatic change and variability can include dehydration, heat or thermal stress and associated decreased growth and productivity as well as death particularly at high temperatures beyond 40°C (Magadza 1994; Hulme 1996). Warming can result in decreased feed consumption in animals and conversely, increased water uptake as the animals have to maintain the balance between heat production and heat loss, and so will reduce their feed consumption in order to reduce heat from metabolism with warming (Moreki 2008). Studies on broiler (chicken) production in Botswana have demonstrated that feed consumption in the birds is reduced by five per cent for every 1°C rise in temperature between 32-38°C (Moreki 2008). Evidence suggests a strong nexus between ecosystems, energy, food production, livelihoods, and climate change across the continent (Hulme 1996; Hulme et al. 2001; IPCC 2007; Magadza 1994). Droughts in particular, whose frequency has increased across the arid to semi-arid ecological zones of the east and southern African region (Hewitson and Crane 2006; Boko et al. 2007; Shongwe et al. 2009; Shongwe et al. 2011) are causing a general trend of reduced rainfall and therefore water stress for plant and animal biomass production (Magadza 1994; Beck and Bernauer 2011), decreased seepage to groundwater stores, and reduced river inflows in these regions (Beck and Bernauer 2010; Hecky et al. 2010; Ndebele-Murisa et al. 2010; Beck and Bernauer 2011; Haande and et al. 2011).

Climate change impacts are now being observed in all African freshwaters, especially in the Zambezi River Basin (Verburg et al. 2003; Tierney et al. 2010; Ndebele-Murisa et al. 2011b; Ndebele-Murisa et al. 2013; Ndebele-Murisa et al. 2014), further devastating these fragile ecosystems by disrupting their ecological functioning. Significant changes in regional climate trends also have major impacts on livelihoods and food security (UNEP 2011). Flooding and droughts are occurring in alternation in the east and southern African region, causing damage to both human livelihoods and river as well as lake habitat integrity, and to some extent, debilitating conservation efforts (Hoeinghaus et al. 2007; Teresa and Casatti 2012). Climate has caused reduction of mixing events, increased thermal stability and reduced nutrients availability in the photic zones of most lakes in the region; thus resulting in decreased primary productivity, reduction in fish sizes, and causing poor fish catches—potentially inducing loss of fish biodiversity (O'Reilly et al. 2003; Ndebele-Murisa et al. 2011a; Ndebele-Murisa et al. 2011b). The coupling of the natural including climate change impacts with anthropogenic stressors such as deforestation, unsustainable agriculture, river impoundments, nutrients and sediment loading, synergistically cause ecological changes. And if environmental and resource sustainability are to be achieved, there is a need to seriously consider and actually maintain habitat integrity in the Zambezi River Basin, and to conserve not only the ecological but also genetic diversity in and around the basin (Hoeinghaus et al. 2007; Teresa and Casatti 2012).

Most tropical aquatic ecosystems are sensitive and vulnerable to climate changes. Climate change and warming in these systems and particularly in the Zambezi Basin waters is affecting stratification, thermal dynamics, mixing and hydrodynamics, primary productivity and fisheries production (Hulme et al. 2001; O'Reilly et al. 2003; Vollmer et al. 2005; Verburg and Hecky 2009; Vincent 2010; MacIntyre 2012; Ndebele-Murisa et al. 2013; IPCC 2014). Climate change also causes both drought and flooding by affecting precipitation and rise in both air and water temperatures, which may result in water resource use conflicts in the Zambezi River Basin (Mhlanga 2001; Holmes 2010; Magadza 2010; MoW 2013a). The Zambezi River Basin is prone to cyclic flood and drought events which have significant impacts on food security, and livestock and wildlife in the basin (Tumbare 2008). For a detailed analysis of the climate around the Zambezi Basin, please see Chapter 5 of this book.

In Africa, agriculture is the most important economic sector, accounting for more than 40 per cent of total export earnings (FAO 2004). In Sub-Saharan Africa, its share in total export revenues averages about 70 per cent. This provides impetus for this chapter to dwell briefly on impacts of climate change on agriculture, which remains the backbone of livelihoods across the continent. In addition, agriculture also forms part of the biological diversity of a region particularly genetic variability which is essential for conservation purposes. And with communities relying heavily on natural ecosystems for sustenance as a result of decreased rain-fed agricultural production systems across the continent, the inspection of climate change impacts on livelihoods is justified. Agriculture and agro-ecological systems in general are most vulnerable to climate change, especially in Africa. Food production has been on the decline for most of Sub-Saharan Africa and has not kept pace with the population increase (Clover 2003). Countries such as Lesotho, Malawi, Zambia and Zimbabwe declared a state of disaster because of food shortage in 2002 (Clover 2003). The 2015/2016 was declared an El Niño drought year across southern Africa. This has opposing effects of increased flooding in East Africa with potential negative implications on the region's agricultural production for the season. Over the past 30 years, the area of agricultural lands has increased (from 166 million ha in 1970 to 202 million ha in 1999) at a great cost to the environment. However, these efforts have been absorbed by rapid population growth (FAO 2004).

The main effect of climate change on semi-arid or tropical agro-ecological systems is felt with a significant reduction in crop yields, which may well force large areas of marginal agriculture out of production (Ndebele-Murisa and Mubaya 2015), not only in the east and southern African region but across the continent (Mendelsohn et al. 2000; FAO 2004). The continent will very likely be on the negative of crop production, with net losses of up to 12 per cent of the region's current production (Parry et al. 1999; Gitay et al. 2001). It is also estimated that up to 40 per cent of Sub-Saharan African countries will lose a rather substantial share of their agricultural resources (implying a loss of US$10-60 billion at the 1990 prices) (FAO 2004). The distribution of these losses is not uniform as certain countries will be affected more than others. For instance, it is projected that by 2100, Chad, Niger and Zambia will lose practically their entire farming sector (Mendelsohn et al. 2000). Similarly, in Zimbabwe, the five agro-ecological zones which were demarcated in the 1960s based on rainfall and soil features have shifted significantly towards aridity; with implications for the type of both domestic farming and wildlife activities that can and can no longer be practiced efficiently in these zones (Ndebele-Murisa and Mubaya 2015).

Local Pressures and Drivers of Ecological Changes in the Zambezi River Basin

The Zambezi River Basin has abundant water, fertile land and soils for agriculture and diverse habitats that are home to large populations of wildlife (SARDC et al. 2012). This natural capital defines the basin's economic activities which include agriculture, forestry, manufacturing, mining, conservation and tourism in the river basin. The basin has witnessed a drastic change in its natural environment in recent years, mainly as a result of demographic dynamics, and urbanization and increasing demand for agricultural land (see Lambin et al. 2003; Nindi 2007). These drivers have brought about changes to ecosystems, water resources and the way different cultures interact. There are several potential threats to the ecology and water quality of the basin. Most of the pressures and threats are related to human impacts; such as high human population growth rates, increasing urbanization within and around the basin, potentially unsustainable agricultural expansion and irrigation, hydropower production demands, deforestation, untreated wastewater discharges, oil and gas exploration and exploitation, mineral exploration and mining in the various catchments, as well as uncontrolled water abstractions as discussed earlier.

While high population growth rates tend to increase deforestation caused by new settlement areas and expansion of farmlands, urbanization increase the demand for portable water and untreated wastewater discharges that deteriorate the quality of water (Berg et al. 1995; Odada et al. 2003; Mhlanga et al. 2013). SARDC et al. (2012) identified climate change and population growth as major drivers of environmental change in the basin. For instance, the population of the basin in 1998 was 31.7 million compared to 40 million in 2008, with 7.5 million people living in urban centres (SARDC et al. 2012). Most countries in the basin are urbanizing rapidly, exerting extra pressure on its finite resources. Neighbouring countries often face similar challenges related to environmental change in a shared natural area and the impacts are mostly felt by people and on livelihoods.

Soil erosion, which is a direct result of deforestation and unsustainable agricultural activities, increases the input of sediments, nutrients, metal pollution and organic matter into rivers and lake ecosystems (Cohen et al. 1993; O'Reilly 1998; Kidd et al. 1999; Hecky et al. 2003; MoW 2013c). It is no exception as a local pressure and driver of ecosystem changes in the Zambezi River Basin. This then causes changes in light attenuation in aquatic habitats and may also cause reduction in species diversity and loss of genetic and ecological differentiation among cichlid fish species (Seehausen et al. 1997; Balirwa et al. 2003; Castillo Cajas et al. 2012). These precious fish are a biodiversity treasure to the Great lakes of Africa (Bootsma and Jorgensen 2004). In the east and southern African region, Cabora Bassa, Chilwa, George, Kariba, Malawi, Malombe, Manyara, Naivasha, Natron, Tanganyika, Tana, Turkana and Victoria constitute the major lakes in the African Rift Valley as well as the Zambezi River Basin and herein, as described earlier, many ecological changes have been noted.

The use of pesticides, herbicides and other agrochemicals in the Zambezi River Basin often has far-reaching consequences since these chemicals can accumulate up the food chain, becoming toxic to organisms, especially top predators (Kidd et al. 2001, Mziray and Kimirei 2016) where humans are positioned. The effects of using dichloro-diphenyl-trichloroethane (DDT) to control the tsetse fly and mosquitoes have been identified with traces of the chemical detected in crocodiles (Phelps and Billing 1972; Berg et al. 1995), lactating mothers (Magadza 1995), and as causing reproductive failure in cormorants (birds) in and around Lake Kariba (Douthwaite 1992). In addition, DDT is thought to have caused lack of sexually active males among the synodontid fish species, *Synodontis zambezensis*, in the same lake. While a sex ratio of 42.6 per cent males to 57.4 per cent females is common among the juvenile stages of the fish species, the post-juvenile stages are almost devoid of sexually active males, with an average of 18.5 per cent females showing sexual maturity, compared to only 2.1 per cent males in this lake (Sanyanga 1996). It has been reported that 86 per cent of 5.2 million hectares of the land area that is cultivated annually in the Zambezi River Basin-which belong to Malawi, Zambia and Zimbabwe together use fertilizers and agrochemicals that are contributing to the growth of harmful aquatic plants (SARDC et al. 2012).

Table 3.1: Some characteristics of and documented ecological changes in four lakes in east and southern Africa

Attributes	Natural/man-made Lakes				References
	Victoria	Tanganyika	Nyasa/Malawi	Kariba	
Max depth (m)	80	1470	700	97	(Balon and Coche, 1974, Coulter, 1991, Lehman, 2009)
Mean depth (m)	40	570	292	29	(Balon and Coche, 1974, Coulter, 1991, Lehman, 2009)
Trophic state	Eutrophic	Oligotrophic	Oligotrophic	Oligo-mesotrophic	(Coulter, 1991, Mahere et al., 2014, Ndebele-Murisa et al., 2014)
Fish Biodiversity/ species	600+	400+	1000+	41	(Konings, 2007, Marshall, 2012, Salzburger et al., 2005, Salzburger and Meyer, 2004)
Population growth	H	H	H	H	(Magadza, 1996, Sitoki et al., 2010, Mölsä et al., 1999)
Stability	+	++	++	+	(Magadza, 2011, O'Reilly et al., 2003, Verburg and Hecky, 2009, Hecky et al., 1994, Ndebele-Murisa et al., 2014)
Surface temperature	++	++	++	++	(Chifamba, 2000, Kraemer et al., 2015, Ndebele-Murisa et al., 2013a, O'Reilly et al., 2003)
Deep water temperature	++	++	+	+	(Magadza, 2011, Mahere et al., 2014, Ndebele-Murisa et al., 2014, O'Reilly et al., 2003, Sitoki et al., 2010, Vollmer et al., 2005)
Mixing events	--	--	--	-	(Magadza, 2010, Mahere et al., 2014, O'Reilly et al., 2003, Sitoki et al., 2010, Verburg and Hecky, 2009, Vollmer et al., 2005, MacIntyre, 2012, Ndebele-Murisa et al., 2014)
Euphotic zone	--	++	+	-	(Descy et al., 2010, Mahere et al., 2014, Verburg and Hecky, 2009a, Ndebele-Murisa et al., 2014)
Transparency	--	++	++	-	(Mahere et al., 2014, Odada et al., 2003, Sitoki et al., 2010, Verburg and Hecky, 2009, Vollmer et al., 2005, Ndebele-Murisa et al., 2014)
Phytoplankton community	Δ	Δ	Δ*	Δ**	(Cronberg, 1997, Hecky et al., 1994, Hecky et al., 1999, Ndebele-Murisa et al., 2011a, Sitoki et al., 2010)
Algal biomass	++	--	--	-	(O'Reilly et al., 2003, Silsbe et al., 2006, Sitoki et al., 2010, Hecky, 1993, Hecky et al., 1994, Ndebele-Murisa et al., 2014)
Zooplankton community	Δ	Δ	Δ	Δ	(Kurki et al., 1999, Mahere, 2012, Masundire, 1997, Ndebele-Murisa et al., 2011a)

Natural/man-made Lakes

Attributes	Victoria	Tanganyika	Nyasa/Malawi	Kariba	References
Zooplankton biomass	Δ	--	--	Δ	(Masundire, 1997, Sarvala et al., 1999, Lehman, 2009)
Primary productivity	++	--	--	-	(Bergamino et al., 2010, O'Reilly et al., 2003, Ndebele-Murisa et al., 2014, Hecky, 1993, Hecky et al., 1994)
Attributes	Victoria	Tanganyika	Nyasa/Malawi	Kariba	References
Fisheries production	--	--	--	--	(Kolding et al., 2003, Mahere, 2012, Ndebele-Murisa et al., 2013a, Sarvala et al., 2006, Tweddle et al., 2015, Banda et al., 2005)
Fishing effort	++	++	++	++	(Mahere, 2012, Ndebele-Murisa et al., 2011a, Van der Knaap et al., 2014b, Kayanda et al., 2009, Van der Knaap et al., 2014a, Banda et al., 2005)
Illegal fishing	++	+	++	++	(Mahere, 2012, Ndebele-Murisa et al., 2011a, Ribbink, 2001, Van der Knaap et al., 2014a, Van der Knaap et al., 2014b, Tweddle et al., 2015, LKFRS, 2010, Marshall, 2012)
Exotic species	++	+	+	++	(Chifamba and Videler, 2014, Kudhongania et al., 1992, Ogutu-Ohwayo, 1990, Witte et al., 2009, Witte et al., 1992, Marufu et al., 2014)
Habitat modification	++	+	++	+	(Chakona et al., 2008, Odada et al., 2003, Weyl et al., 2010, Phiri et al., 2012)
Sedimentation	++	++	++	++	(Alin et al., 1999, 2002, Hecky et al., 2003, Kunz et al., 2011)
Land use	Δ	Δ	Δ	Δ	(Alin et al., 1999, Hecky et al., 2003, William, 2009, Tumbare, 2008, Mhlanga et al., 2014)
Climate change	Δ	Δ	Δ	Δ	(Kraemer et al., 2015, Ndebele-Murisa et al., 2011a, O'Reilly et al., 2003, Marshall et al., 2013, Sitoki et al., 2010, Vollmer et al., 2005)
Livelihood impact	--	--	--	--	(Mölsä et al., 1999, Nyikahadzoi and Raakjaer, 2014, Odada et al., 2003, Weyl et al., 2010)
Chemical pollution	++	nd	++	++	(Campbell et al., 2003, Odada et al., 2003, Shoko and Love, 2005)

H = High; ++ = significant increase, + = increase, -- = significant decrease, - = decrease, Δ = changed significantly, ± = 50/50, nd = no data, * = changes in the southern part of the lake, ** = changes recorded in the Sanyati Basin of the lake

Mining is another source of pressure to the ecosystems in the Zambezi River Basin (Ashton et al. 2001). As the countries in the Zambezi River Basin exploit their mineral potential and all exploration licenses are changed into mining ones the inevitable awaits if poor environmental management is invoked, that is increase in atmospheric and alluvial depositions of the various metals and chemicals, some of which are of health concern (Douthwaite 1992; Kidd et al. 1999; Kidd et al. 2001; Chifamba 2007) will escalate. This is because metal pollution in the water systems such as in the Lake Kariba Basin is considered high (Douthwaite 1992; Berg et al. 1995; Chifamba 2007). SARDC et al. (2012) show mining as a major economic activity in the basin and satellite images reveal striking land-use changes as a result of mining activities, notably in Zambia. The revival of copper mining at Kanshanshi and Lumwana mines in Solwezi in north-western Zambia has led to a population influx, resulting in the rapid but haphazard expansion of the town. As a result, surrounding forested areas have been cleared for firewood and peri-urban farming. Sub-basins such as the Luangwa River, Lake Kariba, and the Kafue and Kabompo rivers have high concentrations of mining operations, contributing to water pollution in the Zambezi River. In addition, highly urbanized sub-basins such as the Kafue and Manyame are discharging untreated or partially treated waste directly into the Zambezi River system (Mwedzi et al. 2016).

Fishing pressure and unsustainable fishing practices are also exerting pressure on fish resources in the Zambezi River Basin. As a result of high fishing pressure, fish catches have declined in the major and minor waters of the basin (Ogutu-Ohwayo 1990; Kudhongania et al. 1992; Turner 1995; Kolding et al. 2003; O'Reilly et al. 2003; Sarvala et al. 2006; Troell and Berg 2008; Ndebele-Murisa et al. 2013). The Chambo fishery in southern Lake Nyasa/Malawi, for instance, has been negatively impacted by unsustainable gillnetting (Banda et al. 2005; Banda 2009). Both fishing pressure and climate change have synergistically caused declines and fluctuation in fish catches in the Zambezi River Basin (O'Reilly et al. 2003; Bulirani 2005, Chitamweba and Kimirei 2005; Kimirei and Mgaya 2007; Plisnier et al. 2009, Ndebele-Murisa et al. 2011a; Ndebele-Murisa et al. 2011b; Loiselle et al. 2014); which jeopardizes community livelihoods and food security.

Livelihood Perspectives: A Case of the Manyame Catchment

The threats posed by potential impacts of environmental and ecological changes on livelihoods dependent on availability of natural resources, e.g. farming, fishing and herding, are compounded by demographic pressures. This brings about a need to analyze trends in ecology and natural resources in the Zambezi River Basin. In developing an in-depth understanding on the perceptions of the men and women in the Zambezi Basin with regard to ecological changes, we place centrality on the livelihoods approach (see Chapter 2 of this volume). Several authors (Scoones 2009; Mubaya 2010) noted that livelihood perspectives have been central to rural development thinking and practice in the past decade. The current case study defines a livelihood as comprising the capabilities assets (including both material and social resources) and activities for a means of living. A livelihood is sustainable when it can cope with and recover from stresses and shocks and maintain or enhance its capabilities and assets, while not undermining the natural resource base (Scoones 2009). We present a case study of the perceptions on ecological changes in the Manyame River basin (see Box 3.1). A questionnaire was administered to 150 households within communities settled along the Manyame River in the Raffingora Area in November 2014 to capture perceptions regarding ecological changes, status and resource use. The Raffingora area is mostly used for commercial farming and is surrounded by mines.

Community experiences of livelihoods influenced how we conceptualized what they meant with regards to the importance of the available natural resource respondents (Table 3.2). We highlight perceptions in the case study that the least beneficial resource in the basin is land (Figure 3.2), which according to SARDC et al (2012) can be attributed to the increase in the human population of the Zambezi Basin which rose by about 25 per cent from 31.7 million in 1998, to about 40 million a decade later. This represents a high population growth rate of ~2.5 per cent per annum.

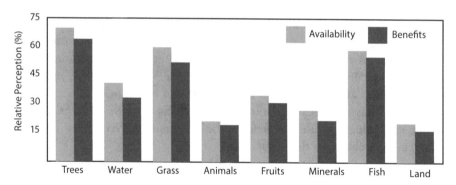

Figure 3.2: Relative perceptions of respondents on availability and benefits of natural resources

As a result of the pressure on land, rain-fed agriculture has been replaced by mining as a major economic activity in the Manyame catchment. Serious land use changes have led to the reduction in physical space in the Zambezi River Basin (SARDC et al. 2012). The finding that fish is a major resource within the Manyame is interesting given that traditionally, fish has not been a major protein source in Zimbabwe (Sen 1995). The implication is that there is need for deliberate efforts to promote fish as a requisite source of protein along its value chain. It is also important to consider that fish is a naturally occurring aquatic resource which does not deplete easily if harvested sustainably (FAO 2007). Females tend to be the custodians of food in homes while males tend to be inclined towards natural resources such as trees and forests (Figure 3.2, Table 3.2) given the commercial role that these resources play in construction and the leading role they play in building processes in the home (Rodd 1991; Wenz 2001). This could explain the inclination by females towards conceptualizing fish as a critical resource and males' preference skewed towards trees and forests. However, it is a cause for concern that perceptions indicate a decline in forest resources. Similarly, a high demand for firewood has seen the deforestation of large areas in countries such as Malawi, Mozambique, Zambia and Zimbabwe especially with the highest number of people living in the rural areas.

Deforestation is also done to clear land for subsistence farming and, more recently, tobacco curing in these same areas (SARDC et al. 2012). Essentially, there is an indication that livelihoods no longer rely on land but on other natural resources such as fish. Despite changes in availability, benefits and time, communities still rely heavily on other natural resources more than land. In terms of gender analysis, males and females prioritize natural resources differently based on their daily interactions with their natural resources.

In contrast to the community in Raffingora/Manyame who reported not noting changes in the weather, communities in Monze and Sinazongwe in the part of southern Zambia, which is within the Zambezi Basin, reported being aware of climate changes and variability. These communities have adjusted their farming systems to cope with these changes such as changing planting dates and using drought-resistant varieties (Mubaya et al. 2012; Ndebele-Murisa and Mubaya 2015). The same communities identified a multiplicity of stressors in addition to climate change including, limited access to credit and agricultural inputs, and reduced household food availability, among others. They however ranked climate change as the major stressor.

Implications for Ecosystem Goods and Services

Aquatic ecosystems are valuable resources, which provide goods and services that underpin and sustain livelihoods and economies of multitudes of communities, societies and nations living around and within these ecosystems. The ecosystem goods and services, which refer to the values and benefits that humans obtain from ecosystems, are therefore fundamental for human well-being (Millennium Ecosystem Assessment 2005). Both rivers and lakes provide the cheapest and easily accessible source of protein in terms of fish in the east and southern Africa region. Studies estimated that fisheries sector supports over one million people in the Lake Tanganyika Basin (Mölsä et al. 1999), about 1.6 million in Malawi (GoM 2011), more than 35 million around Lake Victoria (Weston 2015) and about 300,000 around Lake Kariba (Thieme et al. 2013) and 400,000 to 700,000 in the Cabora Bassa basin (Beilfuss 1999). Moreover, these basins serve as source of portable water by communities, water for mining and other industries, hydropower generation, for livestock and agricultural activities (Euroconsult Mott MacDonald 2008; World Bank 2010; Beilfuss 2012).

The Zambezi River has a very high hydropower potential with, for example, the Shire River in Malawi currently providing over 95 per cent of electricity to the country (Chafota et al. 2005); while Cabora Bassa is the main electricity source for Mozambique—producing over 2, 000 MW (Beilfuss 2012). Currently, Kariba has a potential power output of 1,080 MW. However, with the expansion of the Kariba south power station and an additional 2,400 MW from the proposed Batoka project, the Kariba dam is now producing more than half of the installed hydropower capacity. The entire Zambezi River Basin has an installed capacity of 5,000 MW and an additional planned 13,000 MW (World Bank 2010; Beilfuss 2012). In addition to energy provision, irrigated agriculture in the east and southern African region is projected to increase and thus will benefit immensely from ecosystems such as the Zambezi River, thereby ensuring food and energy security to the region's fast-growing population (Euroconsult Mott MacDonald 2008; World Bank 2010). As such, energy, water and food security in the region hinges largely on proper functioning of aquatic ecosystems, thereby defining the critical need to conserve them. Lakes and rivers such as the Zambezi are also used for navigation; transporting goods and services and businesses across villages, towns, cities and national boundaries.

Box 3.1: Perceptions of availability and benefits from natural resources within the Manyame River catchment, Zimbabwe

Community perceptions on availability and benefit of various natural resources in the Manyame catchment indicate that trees (61 per cent), fish (56 per cent) and grass (53 per cent) are the most beneficial resources and that benefit is largely determined by availability (Figure 3.2). There is, however, a gender dichotomy on some of the livelihood sources where males and females and the married and widowed had different perceptions on the benefits of, for example, trees and fish (see Table 3.2). While trees were considered most beneficial by males, fish were perceived as such by females. There is a general agreement, however, that livelihoods no longer centre mainly on land and animals but on other natural resources such as fish and trees (Figure 3.2; Table 3.2). Essentially, livelihood systems are shifting from the traditional and predictable ones to other livelihood sources that were not common before in this community. The shift is mainly caused by a decrease in most of these resources. For example, 100 per cent of both males and female respondents respectively, perceive land benefits to have decreased the most. All resources except fish were perceived by both males and females alike to have decreased over time. Over 50 per cent of respondents (males and females) reported that water, wood, fruits, grass and animals have declined significantly; while fish resources were perceived by 50 per cent of the respondents to have increased. However, gender differences in perceptions of the declined resources were observed, only for water and animals by males (>50 per cent) and fruits for females (58 per cent). By implication, there is a general perception of change in the ecosystem within the community skewed towards a decline for most of the resources except fish.

Wood and minerals appear to have maintained their role as very important resources in the livelihoods of this Manyame River Basin community over the past five years. The same trends have been maintained for grass except that males place higher priority on wood as compared to females. Females currently place a higher priority on minerals more than five years ago. Generally, females tend to place a lower priority on grass, fruit/trees and animals compared to their male counterparts. There appears to be no significant change in priority of resource for both males and females regardless of the changes in availability and benefits as identified in the previous sections. Generally, females than males perceived crop yields, crop types, crop pests/ diseases, livestock populations, livestock diseases, soil erosion, water erosion, wind erosion and food availability to have increased substantially. With males having a general perception that water availability, rainfall amounts, quality of pastures and income from agriculture have increased more than females did. The majority (>50 per cent) of both males and females respondents reported to have noticed changes in rainfall and temperature, and population increase and macro-economic conditions in the river basin. A greater percentage (55 per cent) of the respondents said that they did notice a significant change in weather in the river basin (Source: Mubaya, unpublished data).

Furthermore, aquatic ecosystems provide regulatory functions in which they support the environment and other ecosystems at large. The regulatory function provided by aquatic ecosystems includes carbon dioxide sequestration, erosion control, climate regulation, water purification, nutrient and water cycling, and sediment and pollutant trapping. The regulatory services sustain ecosystem functioning, hence ecosystem services. Aquatic ecosystems also serve as critical habitats for fish and other aquatic organisms. Lake Nyasa/Malawi and Tanganyika, for example, host over 700 and 400 species of cichlids, respectively, which fetch foreign currency for the riparian countries through tourism, fisheries and aquarium fish trade (Turner et al. 2001; Konings 2007). Most rivers also serve as refuges, nurseries and or breeding grounds for numerous fish, especially potamodromous, anadromous and catadromous species (Tweddle 1983; Tweddle 2001; Chale 2010). Other ecosystem services accrued from rivers such as the Zambezi and lakes include invaluable aesthetic (tourism) and cultural values that are often difficult to quantify or cost (Chafota et al. 2005; Mutasa and Ndebele-Murisa 2015).

Table 3.2: Perceptions of income sources from natural resources expressed across gender and marital status

Income	Male	Female	Married	Widowed	Polygamist
Trees	70.9	20	70.4	20	100
Water	30.9	60	31.5	60	0
Grass	54.5	20	53.7	20	100
Animals	20.0	20	20.4	20	0
Fruits	34.5	20	35.2	20	0
Minerals	21.8	40	22.2	40	0
Fish	54.5	80	53.7	80	100
Land	12.7	20	13.0	20	0

The ecological changes in the aquatic ecosystems of east and southern Africa are affecting provision of ecosystem goods and services. Impairment of provision of ecosystem services is certainly affecting and will continue to affect human well-being in various ways. Reduced water levels, changing productivities of lakes and dwindling fisheries productions in the region have serious implications for food security and economies of the region. For example, fish catches from Lake Tanganyika have declined from about 110,000 tons in 1985 to recent levels of less than 60,000 tons (Tanzania) (Kimirei et al. 2008; Kimirei unpublished data). And while the Chambo fishery in Lake Nyasa/Malawi is almost collapsing, the catches from Lake Kariba have also declined from more than 15,000 tons to less than 5,000 tons between 1994 and 2010 (Ndebele-Murisa et al. 2011a, 2013). Declining fish catches mean hiking fish prices and unavailability of the same to the poor populace. It also means less foreign currency into the economies of the countries within the region, and reduction in the employment attributed to fisheries. Deforestation and unsustainable agriculture in the catchment of both rivers and lakes increase sedimentation rates, thereby negatively affecting breeding and nursery grounds of many fish species. This may result in reduction of several fish species or detrimental hybridization and loss of biodiversity (Ribbink et al. 1983; Tweddle 1983; Seehausen et al. 1997; O'Reilly 1998; Tweddle 2001; Witte et al. 2012). Therefore, the implications of ecological changes on provision of ecosystem goods and services by these fragile aquatic ecosystems and effects on communities' livelihoods cannot be overemphasized.

Ecosystems into which excessive sedimentation and nutrient inputs are rampant can become eutrophic, thereby obstructing access to water or increasing the costs of water production. Also, eutrophic and murky waters may be rendered unsuitable for drinking, bathing, swimming and other recreational activities and therefore negatively affect tourism. A case in point within the Zambezi River Basin is Lake Chivero in Zimbabwe, whose aesthetic value has declined significantly over the years due to its hyper-eutrophic nature and associated anoxic conditions, murky and smelly waters caused by algal scums, floating water hyacinth and fish kills (Moyo 1997; Ndebele 2003; Marufu and Chifamba 2013). Lakes and rivers can also fill up as a result of excessive sediments from catchments that are degraded, thus rendering breeding and nursery habitats unavailable for fish. Sedimentation also reduces the capacity for hydropower production by filling dams with sediments, thereby reducing water levels. The construction of dams, such as the Kariba and Cabora Bassa dams, for hydroelectric power production has been noted to significantly alter the flow regime in the lower Zambezi River Basin (Beilfuss and Dos Santos 2001). For example, higher seasonal variability in flow regime has already been noted; where mean monthly flows/discharges have decreased by about 32 per cent from 7,436 m^3s^{-1} to 2,868 m^3s^{-1} during the high flow (wet) months (March) and from 617 m^3s^{-1} to 2,115 m^3s^{-1} in the low-flow period (November) (Beilfuss and Dos Santos 2001). The changes in the flow regime have affected the downstream ecosystem particularly in the Delta (Mozambique). Natural seasonal flood cycles have been altered with floodplain inundation occurring only during major floods. This has in turn affected and changed the agriculture (flood irrigation to irrigated agriculture), fishery, wildlife, vegetation and livelihood characteristics in the river basin.

Conclusion

Aquatic ecosystems and their immediate environment are important providers of ecosystem goods and services to riparian communities. Their importance cannot be overemphasized, especially in supporting community livelihoods, biodiversity, hydropower production, food security and portable water for domestic use. It is a bitter fact, however, that the ecological settings of many of these ecosystems in the east and southern African region have and are changing. The drivers of these changes are related to both natural forces and human activities. While good management practices will ensure continued provision of the different goods and services to the communities, poor management jeopardizes both the provision of ecological functions and ecosystem goods and services (see Figure 3.1). The human population growth, over-utilization of natural resources—fish, water, forests and minerals; land use change, sedimentation and eutrophication, and alien species' invasions of and introductions into aquatic ecosystems, and climate change form the most important drivers of ecological changes in the region. The changes have and will continue to affect communities that are directly dependent on these systems for food and water and other ecosystem goods and services. It is therefore important to consider some ways in which the effects of the changes can be absorbed or reduced or coped with. For instance, according to community perceptions presented in this chapter, there is an apparent gradual shift of priority resources from land to fish and trees as very important components of rural livelihoods within part of the Zambezi basin. It becomes important for policy making and implementation to promote those resources that are important for these communities' livelihoods in order to deal with reduced production and productivity in ecological systems and small-holder agriculture as well as other emergent forms of livelihood (fisheries, mining) in the region.

One step to achieving this will be to implement proper management measures that interface with appropriate and better land-use practices and planning. The cross ecosystem management approach should be adopted instead of protecting single ecosystems in isolation. Communities should be encouraged to take part in protecting headwaters and maintaining forest and vegetation buffers around all streams, rivers and lakes in the basin. As an example, Section 57(1) of the National Environmental Management Act of 2004, in Tanzania, requires that a buffer zone of 60 m from aquatic ecosystems should be observed by land developers and other resource users. This law should be adopted across the basin and must be put into force unflinchingly. The observance of this and other environmental laws can help maintain healthier river and lake ecosystems, and the goods and services they provide. Moreover, the consumptive human use of streams, rivers and lake waters (e.g., potable water supply for domestic use, livestock, and irrigation) that can significantly reduce freshwater flows into the receiving ecosystems, reservoirs and hydropower installations should be regulated; otherwise they can alter both the hydrological and ecological integrity of these ecosystems.

Mining and irrigated agriculture are developing in the region and environmental issues such as deforestation and overuse of water resources are common practice in these operations. We do not intend to decimate the importance of these activities in supporting economic development and food security. However, they need to be regulated if environmentally acceptable river flows and sediment accretion into rivers, dams and reservoirs and lakes are to be respectively maintained and controlled in the region/basin (see Chapter 7). If we turn a blind eye and overlook their ecosystem values, and let environmental degradation continue unabated, we will be jeopardizing not only conservation outcomes but also provision of ecosystem services that community livelihood heavily depends on. And considering climate change projections for the region (see Chapter 5); we may not be able to sustain our people in terms of food and water supply. Therefore, as the human population continues to increase, demand for food and other ecosystem goods and services subsequently increases, and conservation needs continue to manifest, community livelihood improvement, sustenance and sustainability should be our prime goal. It is by doing so and actually conserving a mosaic of connected and interlinked forests and vegetation in and around our precious Zambezi River Basin and deliberately executing restoration of degraded and fragmented forests and aquatic ecosystems that socio-economic benefits and ecological sustainability, and the ecosystem services provided by the aquatic environment can be accomplished in the region.

References

Alin, S. R., Cohen, A. S., Bills, R., Gashagaza, M. M., Michel, E., Tiercelin, J. & Al, E., 1999, 'Effects of landscape disturbance on animal communities in Lake Tanganyika', *Conservation Biology*, 1017-1033.

Alin, S. R., O'reilly, C. M., Cohen, A. S., Dettman, D. L., Palacios-Fest, M. R. & Mckee, B. A., 2002, 'Effects of land-use change on aquatic biodiversity: A view from the *paleorecord* at Lake Tanganyika, East Africa', Geology, 30, 1143-1146.

Ashton, P. J., Love, D., Mahachi, H. & Dirks, P. H. G. M., 2001, 'An Overview of the Impact of Mining and Mineral Processing Operations on Water Resources and Water Quality in the Zambezi, Limpopo and Olifants Catchments in Southern Africa', Contract Report to the Mining, Minerals and Sustainable Development (Southern Africa) Project, by Report No. ENV-P-C 2001-042. Harare, Zimbabwe: CSIR-Environmentek, Pretoria, South Africa and Geology Department, University of Zimbabwe, xvi + 336 pp.

Balirwa, J. S., Chapman, C. A., Chapman, L. J., Cowx, I. G., Geheb, K., Kaufman, L., Lowe-Mcconnell, R. H., Seehausen, O., Wanink, J. H., Welcomme, R. L. & Witte, F., 2003', 'Biodiversity and Fishery Sustainability in the Lake Victoria Basin: An Unexpected Marriage?', *BioScience*, 53, 703-715.

Balon, E. K. & Coche, A. G. 1974. Lake Kariba. A man-made tropical ecosystem in Central Africa, The Hague, Springer Netherlands.

Banda, M. C., 2009, 'Lake Malawi gillnet fishery management policy brief.Malawi: Ministry of Agriculture and Food Security', Department of Fisheries.

Banda, M. C., Kanyeree, G. Z. & Rusuwa, B. B., 2005, 'The status of the Chambo in Malawi: fisheries and biology', 1-7, in M. C. Banda, D. Jamu, M. Makuwila, & A. Maluwa, eds, The Chambo restoration strategic plan, Mangochi: World Fish Centre.

Beck, L. & Bernauer, T., 2010, 'Water Scenarios for the Zambezi River Basin, 2000-2050', Oslo: International Peace Research Institute, 29 pp.

Beck, L. & Bernauer, T., 2011, 'How will combined changes in water demand and climate affect water availability in the Zambezi River Basin?' *Global Environmental Change*, 21, 1061-1072.

Beilfuss, R., 2012, A risky climate for Southern African Hydro: Assessing hydrological risks and consequences for Zambezi River Basin dams, California, USA: International Rivers, 60 pp.

Beilfuss, R. D. & Dos Santos, D., 2001, 'Patterns of hydrological change in the Zambezi Delta, Mozambique', Working paper #2, Program for the sustainable Management of Cahora Bassa Dam and the Lower Zambezi Valley. Baraboo, USA: The International Crane Foundation, 159pp.

Benedict, C. S., Sudre, C. P., Rodrigues, R., Riva, I. N. & Pereira, M. G., 2007, Multicategoric and qualitative descriptors to estimate the phenotypic variability between pepper access', *Scientia Agraria*, 8, 149-156.

Berg, H., Kiibus, M. & Kautsky, N., 1995, 'Heavy metals in tropical Lake Kariba, Zimbabwe', *Water, Air, and Soil Pollution*, 83, 237-252.

Bergamino, N., Horion, S., Stenuite, S., Cornet, Y., Loiselle, S., Plisnier, P. & Al, E., 2010, 'Spatio-temporal dynamics of phytoplankton and primary production in

Lake Tanganyika using a MODIS based bio-optical time series', *Remote Sensing of Environment*, 114, 772-780.

Boko, M., Niang, I., Nyong, A., Vogel, C., Githeko, A., Medany, M., Osman-Elasha, B., Tabo, R. & Yanda, P., 2007, 'Climate Change 2007: Impacts, Adaptation and Vulnerability. Contribution of Working Group II to the Fourth Assessment Report of the Intergovernmental Panel on Climate Change: Africa', in M. L. Parry, O. F. Canziani, J. P. Palutikof, P. J. Van Der Linden, & C. E. Hanson, eds, IPCC Fourth Assessment Report: Climate Change 2007: Working Group II: Impacts, Adaptation and Vulnerability, Cambridge University Press, pp. 433-467.

Bootsma, H., Mwita, J., Mwichande, B., Hecky, R., Kihedu, J. & Mwambungu, J.,1999, 'The atmospheric deposition of nutrients on Lake Malawi/Nyasa', in H. A. Bootsma & R. E. Hecky, eds, Water Quality Report, Lake Malawi/Nyasa Biodiversity Conservation Project, SADC/GEF, pp. 85-111.

Bootsma, H. A. & Jorgensen, S. E., 2005, 'Lake Malawi: Experience and Lessons Learned Brief' in International Lake Environment Committee Foundation, eds, Lake Basin Management Initiative: Experience and Lessons Learned Briefs (on CD). Kasatsu, Japan, pp. 259-276.

Bulirani, A., 2005, 'Observations on the factors behind the decline of the Chambo in Lake Malawi and Lake Malombe', in M. C. Banda, D. Jamu, F. Njaya, M. Makuwila, & A. Maluwa, eds, The Chambo restoration strategic plan. Mangochi: World Fish Centre.

Calder, I. R., Hall, R. L., Bastable, H. G., Gunston, H. M., Shela, O., Chirwa, A. & K. R., 1995, The impact of land use change on water resources in sub-Saharan Africa: a modeling study of Lake Malawi, *Journal of Hydrology*, 170, 123-135.

Campbell, L. M., Osano, O., Hecky, R. E. & Dixon, D. G., 2003, 'Mercury in fish from three Rift Valley lakes (Turkana, Naivasha, Baringo), Kenya, East Africa', *Environmental Pollution*, 125, 281-286.

Carvalho, D. R., Castro, D., Callisto, M., Moreira, M. Z. & Pompeu, P. S., 2015, 'Isotopic variation in five species of stream fishes under the influence of different land uses', Journal of Fish Biology, 87 (3), 559-578. DOI: 10.1111/jfb.12734.

Castillo Cajas, R. F., Selz, O. M., Ripmeester, E. A., Seehausen, O. & Maan, M. E., 2012, 'Species-specific relationships between water transparency and male coloration within and between two closely related Lake Victoria cichlid species', *International Journal of Evolutionary Biology*, 2012, 161306.

Chafota, J., 2012, 'Integrated natural resources management in a dynamic transboundary watershed context: The Songwe River catchment experience', 121 p: WWF, CDE and ASDC.

Chafota, J., Burgess, N., Thieme, M. & Johnson, S., 2005, 'Lake Malawi/Niassa/Nyasa Ecoregion Conservation Programme: Priority Conservation Areas and Vision for Biodiversity Conservation', WWF-SARPO.

Chakona, A., Phiri, C., Magadza, C. H. D. & Brendonck, L., 2008, 'The influence of habitat structure and flow permanence on macroinvertebrate assemblages in temporary rivers in northwestern Zimbabwe', *Hydrobiologia*, 607, 199-209.

Chale, F. M. M., 2010, 'Preliminary studies on the ecology of Mbasa (Opsaridium microlepis (Gunther)) in Lake Nyasa around the Ruhuhu River', *Journal of Ecology and the Natural Environment*, 3, 58-62.

Chifamba, P. C., 2000, 'The relationship of temperature and hydrological factors to catch per unit effort, condition and size of the freshwater sardine, *Limnothrissa miodon* (Boulenger), in Lake Kariba', Fisheries Research, 45, 271-281.

Chifamba, P. C., 2007, 'Trace metal contamination of water at a solid waste disposal site at Kariba, Zimbabwe', *African Journal of Aquatic Science*, 32, 71-78.

Chifamba, P. C. & Videler, J. J., 2014, 'Growth rates of alien Oreochromis niloticus and indigenous Oreochromis mortimeri in Lake Kariba, Zimbabwe', *African Journal of Aquatic Science*, 39, 167-176.

Chimhowa, A., 2009, Moving forward in Zimbabwe, reducing poverty and promoting growth, The University of Manchester Brooks: The World Poverty Institute.

Chitamweba, D. B. R. & Kimirei, I. A., 2005, 'Present fish catch trends at Kigoma Tanzania', *Verhandlungen Internationale Vereinigung für Limnologie*, 29, 373-376.

Chiuta, T. M. & Johnson, S., 2010, 'Songwe River Transboundary Catchment Management Project (Tanzania-Malawi): Final evaluation report', Swiss Development Cooperation and WWF, 50 pp.

Clover, J., 2003, 'Food security in Sub-Saharan Africa', African Security Review, 12, 5-15.

Cohen, A. S., Bills, R., Cocquyt, C. & Caljon, A. G., 1993, 'The impact of sediment pollution on biodiversity in Lake Tanganyika', *Conservation Biology*, 7, 667-677.

Cohen, A. S., Lezzar, K. E., Cole, J., Dettman, D., Ellis, G. S., Gonneea, M. E., Plisnier, P.-D., Langenberg, V., Blaauw, M. & Zilifi, D., 2006, 'Late Holocene linkages between decade-century scale climate variability and productivity at Lake Tanganyika, Africa', *Journal of Paleolimnology*, 36, 189-209.

Coulter, G. W., 1970, Population changes within a group of fish species in Lake Tanganyika following their exploitation', *Journal of Fish Biology*, 2, 329-353.

Coulter, G. W., 1991, Lake Tanganyika and its life, Natural History Museum Publications, 354 p., London: Oxford University Press.

Cronberg, G., 1997, 'Phytoplankton in Lake Kariba, 1986-1990', in J. Moreau, ed., Advances in the ecology of Lake Kariba, Harare: University of Zimbabwe Publishers.

Crul, R. C. M., 1997, 'Limnology and hydrology of Lakes Tanganyika and Malawi', Studies and reports in hydrology, 54, UNESCO.

Descy, J. P., Hardy, M. A., Stenuite, S., Pirlot, S., Leporcq, B., Kimirei, I., Sekadende, B., Mwaitega, S. R. & Sinyenza, D., 2005, 'Phytoplankton pigments and community composition in Lake Tanganyika', Freshwater Biology, 50, 668-684.

Descy, J. P., Tarbe, A. L., Stenuite, S., Pirlot, S., Stimart, J., Vanderhey-Den, J., Leporcq, B., Stoyneva, M. P., Kimirei, I., Sinyinza, D. & Pli-Snier, P. D., 2010, 'Drivers of phytoplankton diversity in Lake Tanganyika', *Hydrobiologia*, 653, 29-44.

Douthwaite, R. J., 1992, 'Effects of DDT on the Fish Eagle, Haliaeetus vocifer population of Lake Kariba in Zimbabwe, Ibis 134, 250-258.

Euroconsult Mott Macdonald, 2008, 'Integrated Water Resources Management Strategy and Implementation Plan for the Zambezi basin', SADC-WD/Zambezi River Authority, SIDA, DANIDA, Norwegian Embassy Lusaka, 127 pp.

FAO, 2004. The state of insecurity in the world: monitoring progress towards the World Food Summit and Millennium Development Goals. Rome, Italy, 1-43 pp.

FAO, 2007, The State of World Fisheries and Aquaculture, Rome, Italy: FAO Fisheries and Aquaculture Department.

Gitay, H., Brown, S., Easterling, W. & Jallow, B., 2001, 'Ecosystems and their goods and services', in J. J. Mccarthy, O. F. Canziani, N. A Leary, D. J. Dokken, & K. S. White, eds, Climate Change: Impacts, Adaptation, and Vulnerability, Cambridge and New York.

Gom, 2011, Annual Economic Report. Malawi. Lilongwe, Malawi: Ministry of Development Planning and Cooperation.

GoZ, 2008, The Government of Zimbabwe: Zimbabwe Millennium Development Goals 2000-2007. Mid-term Progress Report. Harare, Zimbabwe: UNDP.

Haande, S.Rohrlack, T., Semyalo, R.P., Brettum, P., Edvardsen, B., Lyche-Solheim, A., Sørensen, K. & Larsson, P., 2011, 'Phytoplankton dynamics and cyanobacterial dominance in Murchison Bay of Lake Victoria (Uganda) in relation to environmental conditions', Limnologica, 41 (1), 20-29.

Hara, M., 2011, 'Community response: decline of the Chambo in Lake Malawi's Southeast Arm', in S. Fjentoft, & A. Eide, eds, Poverty mosaics: realities and prospects in small-scale fisheries, Doldretcht: Springer Science+Business Media B. V.

Hecky, R., Mugidde, R., Ramlal, P., Talbot, M. & Kling, G., 2010, 'Multiple stressors cause rapid ecosystem change in Lake Victoria', *Freshwater Biology*, 55, 19-42.

Hecky, R. E., 1993, 'The eutrophication of Lake Victoria', Verhandlungen der Internationalen Vereinigung für theoretische und angewandte Limnologie, 25, 39-48.
Hecky, R. E., Bootsma, H. A. & Kingdon, M. L., 2003, 'Impact of land use on sediment and nutrient yields to Lake Malawi/Nyasa (Africa)', Journal of Great Lakes Research, 29, 139-158.

Hecky, R. E., Bugenyi, F. W. B., Ochhumba, P., Talling, J. F., Mugidde, R., Gophen, M. & Kaufman, L., 1994a, 'Deoxygenation of the Deep Water of Lake Victoria, East Africa', Limnology and Oceanography, 39, 1476-1481.

Hecky, R. E., Kling, H. J., Johnson, T. C., Bootsma, H. A. & Wilkinson, P., 1999, 'Algal and sedimentary evidence for recent changes in the water quality and limnology of Lake Malawi/Nyasa; in H. A. Bootsma & R. E. Hecky, eds, Water Quality Report, Lake Malawi/Nyasa Biodiversity Conservation Project, Washington D.C: SADC (Southern African Development Community) and GEF (Global Environmental Facility).

Hewitson, B. C. & Crane, R. G., 2006, 'Consensus between GCM climate change projections with empirical downscaling: precipitation downscaling over South Africa', International Journal of Climatology, 26, 1315-1337.

Hoeinghaus, D. J., Winemiller, K. O. & Birnbaum, J. S., 2007, 'Local and regional determinants of stream fish assemblage structure: inferences based on taxonomic vs. functional groups', *Journal of Biogeography*, 34, 324-338.

Holmes, N., 2010, 'Threats to river habitats and associated plants and animals', in C. Hurford, M. Schneider, & I. Cowx, eds, Conservation Monitoring in Freshwater Habitats: A Practical Guide and Case Studies, London: Springer Science+Business Media BV, pp. 103-11.

Hulme, D., 1996, Impact Assessment Methodologies for Microfinance: Theory, Experience and Better Practice, University of Manchester, United Kingdom: Institute for Development Policy and Management.

Hulme, M., Doherty, R. M., Ngara, T., New, L. G. & Lister, D., 2001, 'African climate change: 1900-2100', Climate Research, 17, 145-168.

IPCC, 2007, 'Impacts, Adaptation and Vulnerability', in M. L. Parry, O. F. Canziani, J. P. Palutikof, P. J. Van Der Linden, & C. E. Hanson, eds, Contribution of Working Group II to the Fourth Assessment Report of the Intergovernmental Panel on Climate Change, Cambridge, UK: Cambridge University Press.

IPCC, 2014, Climate Change 2014: Synthesis Report. Contribution of Working Groups I, II and III to the Fifth Assessment Report of the Intergovernmental Panel on Climate Change (Core Writing Team, R. K. Pachauri & L. A. Meyer, eds.), IPCC, Geneva, Switzerland, 151 pp.

IUCN, 2010, 'An analysis of the status and distribution of freshwater biodiversity in Continental Africa on the IUCN Red List', http://www.iucnredlist.org/initiatives/freshwater/acknowledgements [Accessed 1 July 2016].

Kamete, A. Y., 2007, 'When livelihoods take a battering. Mapping the 'New Gold Rush' in Zimbabwe's Angwa-Pote Basin', *Transformation: Critical Perspectives on Southern Africa*, 65, 36-67.

Kayanda, R., Taabu, A. M., Tumwebaze, R., Muhoozi, L., Jembe, T., Mlaponi, E. & Nzungi, P., 2009, 'Status of the major commercial fish stocks and proposed species-specific management plans for Lake Victoria', *African Journal of Tropical Limnology and Fisheries*, 12, 15-21.

Kidd, K. A., Bootsma, H. A., Hesslein, R. H., Muir, D. C. G. & Hecky, R. E., 2001, 'Biomagnification of DDT through the Benthic and Pelagic Food Webs of Lake Malawi, East Africa: Importance of Trophic Level and Carbon Source', *Environmental Science and Technology*, 35, 14-20.

Kidd, K. A., Lockhart, W. L., Wilkinson, P. & Muir, D. C. G., 1999, 'Metals, pesticides and other persistent contaminants in water, sediments and biota from Lake Malawi', in H. A. Bootsma, & R. E. Hecky, eds, Water Quality Report, Lake Malawi/Nyasa Biodiversity Conservation Project, SADC/GEF, 243-276.

Kimirei, I. A. & Mgaya, Y. D., 2007, 'Influence of environmental factors on seasonal changes in clupeid catches in the Kigoma area of Lake Tanganyika', *African Journal of Aquatic Science*, 32, 291-298.

Kimirei, I. A., Mgaya, Y. D. & Chande, A. I., 2008, 'Changes in species composition and abundance of commercially important pelagic fish species in Kigoma area, Lake Tanganyika, Tanzania', *Aquatic Ecosystem Health and Management*, 11, 29-35.

Kolding, J., Musando, B. & Songore, N., 2003a, 'Inshore fisheries and fish population changes in Lake Kariba', in E. Jul-Larsen, J. Kolding, J. R. Nielsen, R. Overa, & P. A. M. Van Zwieten, eds, 2003, Management, co-management or no management? Major dilemmas in southern African freshwater fisheries, Part 2: Case studies. FAO Fisheries Technical Paper 426/2, Rome: FAO, pp. 67-99.

Kolding, J., Musando, B. & Songore, N., 2003b, 'Inshore fisheries and fish population changes in Lake Kariba', in E. Jul-Larsen, J. Kolding, J. R. Nielsen, R. Overa, & P. A. M. Van Zwieten, eds Management, co-management or no management? Major dilemmas in southern African freshwater fisheries. Part 2: Case studies. FAO Fisheries Technical Paper 426/2. Rome: FAO.

Konings, A., 2007, Malawi cichlids in their natural habitat, New 4th Edition. Texas: Cichlid Press.

Kraemer, B. M., Hook, S., Huttula, T., Kotilainen, P., O'reilly, C. M. & Peltonen, A., 2015, Century-Long Warming Trends in the Upper Water Column of Lake Tanganyika. *PLoS ONE*, 10, e0132490. doi:10.1371/journal.pone.0132490.

Kudhongania, A. W., Twongo, T. & Ogutu-Ohwayo, R., 1992, 'Impact of Nile perch on the fisheries of Lakes Victoria and Kyoga', *Hydrobiologia*, 232, 1-10.

Kunz, M. J., Anselmetti, F. S., Wuest, A., Wehrli, B., Vollenweider, A., Thuring, S. & Senn, D. B., 2011, 'Sediment accumulation and carbon, nitrogen, and phosphorus deposition in the large tropical reservoir Lake Kariba (Zambia/Zimbabwe)', *Journal of Geophysical Research*, 116, G03003, doi:10.1029/2010JG001538.

Kurki, H., Vuorinen, I., Bosma, E. & Bwebwa, D., 1999, Spatial and temporal changes in copepod zooplankton communities of Lake Tanganyika, *Hydrobiologia*, 407, 105-14.

Lake Kariba Fisheries Research Institute (LKFRI), 2010, Annual Report 2010. Kariba, Zimbabwe: Lake Kariba Fisheries Research Institute (LKFRI).

Leemans, R. & Eickhout, B., 2004, 'Another reason for concern: regional and global impacts on ecosystems for different levels of climate change', *Global Environmental Change*, 14, 219-228.

Lehane, S., 2013, 'Fish for the future: Aquaculture and Food security', Strategic analysis paper, Future Directions International Pvt Ltd, Australia. 8 pp.

Lehman, J. T., 2009, 'Lake Victoria', in H. J. Dumont, ed, The Nile: Origin, Environments, Limnology and Human Use, Dordrecht: Springer Science + Business Media B.V.

LNBWB, 2014, 'Basin annual hydrological report-Nov.2012-Oct.2013', Lake Nyasa Basin Water Board. 59 pp.

Loiselle, S., Cozar, A., Adgo, E., Ballatore, T., Chavula, G., Descy, J. P., Harper, D. M., Kansiime, F., Kimirei, I., Langenberg, V., Ma, R., Sarmento, H. & Odada, E., 2014, 'Decadal trends and common dynamics of the bio-optical and thermal characteristics of the African Great Lakes', PLoS One, 9,e93656.

Mabika, N., Masiya, T., Utete, B., Barson, M. & Tsamba, J., 2015, 'Trace Metal Concentration in Two Matrices in an Urban Subtropical River', *Journal of Water Resource and Protection*, 7, 219-227. doi:10.4236/jwarp.2015.73018.

Macintyre, S., 2012, 'Climatic variability, mixing dynamics, and ecological consequences in the African Great Lakes', in C. R. Goldman, M. Kumagai, & R. D. Robarts, eds, Climatic Change and Global Warming of Inland Waters. Impacts and Mitigation for Ecosystems and Societies, Chichester, UK: John Wiley and Sons Ltd.

Magadza, C. H. D., 1994, 'Climate change: Some likely multiple impacts in southern Africa', *Food Policy* 19,165-191.

Magadza, C. H. D., 1995, 'DDT in the Tropics: A review of mercury the NRI report on impacts of DDT in the Zambezi', Harare, Zimbabwe: Zambezi Society, 31 pp.

Magadza, C. H. D., 1996, 'Kariba Reservoir: Experience and lessons learned', *Lakes & Reservoirs: Research and Management*, 11, 271-286.

Magadza, C. H. D., 2010, 'Indicators of above normal rates of climate change in the Middle Zambezi Valley, Southern Africa', *Lakes and Reservoirs: Research and Management*, 15, 167-192.

Magadza, C. H. D., 2011, 'Indications of the effects of climate change on the pelagic fishery of Lake Kariba, Zambia-Zimbabwe. *Lakes and Reservoirs', Research and Management*, 16, 15-22.

Mahere, T. 2012. A Bioeconomic Assessment of the Kapenta Fishery of Lake Kariba, Zimbabwe. In: ZIMBABWE, U. O. (ed.) Unpublished MSc Thesis. Harare, Zimbabwe: University of Zimbabwe.

Mahere, T. S., Mtsambiwa, M. Z., Chifamba, P. C. & Nhiwatiwa, C., 2014, 'Climate change impact on the limnology of Lake Kariba, Zambia-Zimbabwe', *African Journal of Aquatic Science*, 39 (2), 215-221.

Marshall, B. 2011, The fishes of Zimbabwe and their Biology. Smithiana Monograph 3. South African Institute for Aquatic Biodiversity. Grahamstown.

Marshall, B. E. 2012, Does climate change really explain changes in the fisheries productivity of Lake Kariba (Zambia-Zimbabwe)? *Transactions of the Royal Society of South Africa*, 67, 45-51.

Marshall, B. E., Ezekiel, C. N., Gichuki, J., Mkumbo, O. C., Sitoki, L. & Wanda, F., 2013, 'Has climate change disrupted stratification patterns in Lake Victoria, East Africa?', *African Journal of Aquatic Science*, 38, 249-253.

Marufu, C., Phiri, C. & Nhiwatiwa, T., 2014, 'Invasive Australian crayfish Cherax quadricarinatus in the Sanyati Basin of Lake Kariba: a preliminary survey', *African Journal of Aquatic Science*, 39, 233-236.

Mudimbu, D., Ndebele-Murisa, M.R. & Charakupa-Chingono, T., 2012, 'Promoting social and environmental accountability in southern Africa: Zambezi, Limpopo and Orange Senque Rivers'. Pretoria, South Africa: Southern Africa Resource Watch.

Masundire, H. M., 1997, 'Spatial and temporal variations in the composition and density of crustacean plankton in the five basins of Lake Kariba, Zambia-Zimbabwe', *Journal of Plankton Research*, 19, 43-62.

Mccarty, J. P., 2001', 'Ecological consequences of recent climate change', *Conservation Biology*, 15,320-331.

Mendelsohn, R., Dinar, A. & Dalfelt, A., 2000, Climate Change Impacts on African Agriculture, Preliminary analysis prepared for the World Bank. Washington, D.C: District of Columbia, 25 pp.

Mhlanga, L., 2001, 'Water quality of the Eastern basin of Lake Kariba, Zimbabwe in relation to the encroachment of Eichhornia crassipes', *Verhandlungen der Internationale Vereinigung für theoretische und angewandte Limnologie*, 27, 3595-3598.

Mhlanga, L., Nyikahadzoi, K. & Haller, T., 2013, Fragmentation of Natural Resources Management: Experiences from Lake Kariba, 176 pp, Lit Verlag.

Millennium Ecosystem Assessment, 2005, Ecosystems and Human Well-being: Synthesis, Washington, DC: Island Press, 155 pp.

Mölsä, H., Reynolds, J. E., Coenen, E. J. & Lindquist, O. V., 1999, 'Fisheries research towards resource management on Lake Tanganyika', *Hydrobiologia*, 407, 1-24.

Moon, B., 2011, Secretary-General's remarks to the Security Council on the Impact of Climate Change on International Peace and Security, 20 July 2011, New York.

MOW, 2013a, 'Preparation of an integrated water resources management and development plan for the Lake Nyasa Basin. Climate change report', SMEC,49 pp.

MOW, 2013b, 'Preparation of an integrated water resources management and development plan for the Lake Nyasa Basin. Hydrology report', SMEC,75 pp.

MOW, 2013c, 'Preparation of an integrated water resources management and development plan for the Lake Nyasa Basin. Limnology and fisheries report', Prepared by SMEC, 73 pp.

Msomphora, M. R., 2005, 'Eutrophication of the East African Great Lakes', in P. Wassmann, & K. Olli, eds, Drainage basin nutrient inputs and eutrophication: an integrated approach, pp.279-289, Norway: University of Tromsø.

Mutasa, M. & Ndebele-Murisa, M.R., 2015, 'Biodiversity and the development nexus', in T. Murisa and T. Chikweche, Eds, Beyond the crises: Zimbabwe's transformation and prospects for development, TrustAfrica and Weaver Press, Harare, pp. 198-237.

Mwedzi, T., Bere, T & Mangadze, T., 2016, 'Macroinvertebrate assemblages in agricultural, mining and urban tropical streams: Implications for conservation and management', Environmental Science and Pollution Research, 23 (11), 11181-11192.

Mziray, P. & Kimirei, I. A., 2016, 'Bioaccumulation of heavy metals in marine fishes (Siganus sutor, Lethrinus harak, and Rastrelliger kanagurta) from Dar es Salaam Tanzania', Regional Studies in Marine Science, 7, 72-80.

Nakayama, S. M., Ikenaka, Y., Muzandu, K., Choongo, K., Oroszlany, B., Teraoka, H., Mizuno, N. & Ishizuka, M., 2010, 'Heavy metal accumulation in lake sediments, fish (Oreochromis niloticus and Serranochromis thumbergi), and Crayfish (Cherax quadricarinatus) in Lake Itezhi-tezhi and Lake Kariba, Zambia', Archives of Environmental Contamination and Toxicology, 59, 291-300.

Ndebele-Murisa, M.R. & Mubaya, C.P., 2015, 'The climate change scare: livelihoods and adaptation options for smallholder farmers in Zimbabwe', in T. Murisa, and T. Chikweche, Eds, Beyond the crises: Zimbabwe's transformation and prospects for development, Harare, Zimbabwe, Trust Africa and Weaver Press, pp. 154-197.

Ndebele-Murisa, M., Musil, C. F. & Raitt, L., 2010, 'A review of phytoplankton dynamics in tropical African lakes', South African Journal of Science, 106, 1-6.

Ndebele-Murisa, M. R., Hill, T. & Ramsay, L., 2013, 'Validity of downscaled climate models and the implications of possible future climate change for Lake Kariba's Kapenta fishery', Environmental Development, 5, 109-130.

Ndebele-Murisa, M. R., Mashonjowa, E. & Hill, T., 2011a, 'The decline of Kapenta fish stocks in Lake Kariba-a case of climate changing?', Transactions of the Royal Society of South Africa, 66, 220-223.

Ndebele-Murisa, M. R., Mashonjowa, E. & Hill, T., 2011b, 'The implications of a changing climate on the Kapenta fish stocks of Lake Kariba, Zimbabwe', Transactions of the Royal Society of South Africa, 66, 105-119.

Ndebele-Murisa, M. R., Musil, C. F., Magadza, C. H. D. & Raitt, L., 2014, 'A decline in the depth of the mixed layer and changes in other physical properties of Lake Kariba's water over the past two decades', Hydrobiologia, 721, 185-195.

Ndebele-Murisa, M. R., Musil, F. C. & Raitt, M. L., 2012, 'Phytoplankton biomass and primary production dynamics in Lake Kariba', Lakes and Reservoirs: Research and Management, 17, 275-289.

Ndhlovu, A., 2013, 'The Zambezi: Changing the environment in the Zambezi River Basin, The Zambezi, 8, 1-8.

Nelson, G. C., Bennett, E., Berhe, A. A., Cassman, K., Defries, R., Dietz, T., Dobermann, A., Dobson, A., Janetos, A., Levy, M., Marco, D., Nakicenovic, N., O'Neill, B., Norgaard, R., Petschel-Held, G., D. Ojima, P. Pingali, Watson, R. & Zurek, M., 2006, 'Anthropogenic drivers of ecosystem change: an overview', Ecology and Society, 11, 29. [online] URL: http://www.ecologyandsociety.org/vol11/iss2/art29/.

Nhiwatiwa, T. Barson, M. Utete B, & Cooper, R., 2011, 'Metal concentrations in water, sediments and catfish (Clarias gariepinus, Burchell, 1822) in peri-urban river systems in the Upper Manyame catchment, Zimbabwe', African Journal of Aquatic Sciences, (33)11, 34-41.

Nindi, S. J., 2007, 'Changing livelihoods and the environment along Lake Nyasa, Tanzania', *African Study Monographs*, 36, 71-93.

Njaya, F., Snyder, K. A., Jamu, D., Wilson, J., Howard-Williams, C., Allison, E. H. & Andrew, N. L., 2011, 'The natural history and fisheries ecology of Lake Chilwa, southern Malawi', *Journal of Great Lakes Research*, 37, Supplement 1, 15-25.

Nyikahadzoi, K. & Raakjaer, J., 2014, 'The myths and realities in the management of the Kapenta fishery at Lake Kariba (Zimbabwe)', *Lakes and Reservoirs: Research and Management*, 19, 1-10.

O'Reilly, C.M., 1998 Benthic algal productivity in Lake Tanganyika and the effects of deforestation, Pollution Control and Other Measures to Protect Biodiversity in Lake Tanganyika. Lake Tangayika Biodiversity Project, Tanzania, pp. 1-17.

O'Reilly, C. M., Allin, S. R., Plisnier, P. D., Cohen, A. S. & Mckee, B. A., 2003, 'Climate change decreases aquatic ecosystem productivity of Lake Tanganyika, Africa', *Nature*, 424, 766-768.

Odada, E. O., Olago, D. O., Bugenyi, F., Kulindwa, K., Karimumryango, J., West, K., Ntiba, M., Wandiga, S., Aloo-Obudho, P. & Achola, P., 2003, 'Environmental assessment of the East African Rift Valley Lakes', *Aquatic Sciences-Research Across Boundaries*, 65, 254-271.

Ogutu-Ohwayo, R., 1990, The decline of the native fishes of Lakes Victoria and Kyoga (East Africa) and the impact of introduced species, especially the Nile perch, Lates niloticus, and the Nile tilapia, Oreochromis niloticus', *Environmental Biology of Fishes*, 27, 81-96.

Ogutu-Ohwayo, R. & Balirwa, J. S., 2006, 'Management challenges of freshwater fisheries in Africa', *Lakes and Reservoirs: Research & Management*, 11, 215-226.

Parry, M. L., Fischer, C., Livermore, M., Rosenzweig, C. & Iglesias, A., 1999, 'Climate Change and World Food Security: A New Assessment', *Global Environmental Change*, 9, 51-67.

Phelps, R. J. & Billing, K. J., 1972, 'Pesticide levels from animals in Rhodesia', *Transactions of the Rhodesian Scientific Association*, 55, 6-9.

Plisnier, P. D., Mgana, H., Kimirei, I., Chande, A., Makasa, L., Chimanga, J., Zulu, F., Cocquyt, C., Horion, S., Bergamino, N., Naithani, J., Deleersnijder, E., André, L., Descy, J. P. & Cornet, Y., 2009, 'Limnological variability and pelagic fish abundance (*Stolothrissa tanganicae and Lates stappersii*) in Lake Tanganyika', *Hydrobiologia*, 625, 117-134.

Ramlal, P. S., Hecky, R. E., Bootsma, H. A., Schiff, S. L. & Kingdon, M. J., 2003, 'Sources and Fluxes of Organic Carbon in Lake Malawi/Nyasa', *Journal of Great Lakes Research*, 29, Supplement 2, 107-120.

Ribbink, A. J., 2001, 'Lake Malawi/Niassa/Nyassa Ecoregion based conservation programme: Biophysical reconnaissance', Harare, Zimbabwe. WWF Southern African Regional Programme Office.

Ribbink, A. J., Marsh, B. A., Marsh, A. C., Ribbink, A. C. & Sharp, B. J., 1983, 'A preliminary survey of the cichlid fishes of rocky habitats in Lake Malawi', *South African Journal of Zoology*, 18, 147-310.

Rodd, A., 1991, Women and the Environment, New Jersey, Zed Books Ltd.

Rosenblatt, R. A., 2005, 'Ecological change and the future of the human species: can physicians make a difference?', *Annals of Family Medicine*, 3, 173-6.

Salzburger, W., Mack, T., Verheyen, E. & Meyer, A., 2005, 'Out of Tanganyika: Genesis, explosive speciation, key-innovations and phylogeography of the haplochromine cichlid fishes', Evolutionary Biology, 5.

Salzburger, W. & Meyer, A., 2004, 'The species flocks of East African *cichlid* fishes: Recent advances in molecular phylogenetics and population genetics', *Naturwissenschaften*, 91, 277-290.

Sanyanga, R., 1996, 'The inshore fish populations of Lake Kariba with reference to the biology of *Synodontis zambezensis* Peters, 1852', Doctoral thesis, Stockholm University.

SARDC, SADC, ZAMCOM, GRID-ARENDAL & UNEP, 2012, Zambezi River Basin: Atlas of the changing environment, 148 pp.

Sarvala, J., Langenberg, V. T., Salomen, K., Chitamwebwa, D., Coulter, G. W., Huttula, T., Kanyaru, R., Kotilainen, P., Makasa, L., Mulimbwa, N. & Mölsä, H., 2006, 'Fish catches from Lake Tanganyika mainly reflect changes in fishery practices, not climate', *Verhandlungen Internationalen Vereinigung für Limnologie*, 29,1182-1188.

Sarvala, J., Salonen, K., Jarvinen, M., Aro, E., Huttula, T., Kotilainen, P., Kurki, H., Langenberg, V., Mannini, P., Peltonen, A., Plisnier, P. D., Vuorinen, I., Molsa, H. & Lindquist, O. V., 1999, 'Trophic structure of Lake Tanganyika: Carbon flows in the pelagic food web', *Hydrobiologica*, 407, 149-173.

Scoones, I., 2009, 'Livelihoods Perspectives and Rural Development', *Journal of Peasant studies*, 36(3), 171-196.

Seehausen, O., Van Alphen, J. J. M. & Witte, F., 1997, 'Cichlid fish diversity threatened by eutrophication that curbs sexual selection. *Science*, 277, 1808-1811.

Sen, S., 1995, 'The market for fish and fish products in Zimbabwe', in Aquaculture for Local Community Development Programme, A., ed., Zambia-Zimbabwe SADC Fisheries Project, (Lake Kariba). ALCOM Field Document No. 34.

Shoko, D. S. M. 2002. Small-scale mining and alluvial gold panning within the Zambezi Basin: an ecological time bomb and tinderbox for future conflicts among riparian states. Paper presented at 'The Commons in an Age of Globalization', the Ninth Conference of the International Association for the Study of Common Property, (June 17-21). Victoria Falls, Zimbabwe.

Shoko, D. S. M. & Love, D., 2005, 'Gold panning legislation in Zimbabwe-What potentials for sustainable management of river resources', in K. Mathew & I. Nhapi, eds, Water and Wastewater Management for Development Countries: IWA Water and Environmental Management Series, London: IWA Publishing.

Shongwe, M., Van Oldenborgh, G., Van Den Hurk, B., De Boer, B., Coelho, C. & Van Aalst, M. 2009. Projected changes in mean and extreme precipitation in Africa under global warming. Part I: *Southern Africa Journal of Climate*, 22, 3819-3837.

Shongwe, M., Van Oldenborgh, G., Van Den Hurk, B. & Van Aalst, M., 2011, 'Projected changes in mean and extreme precipitation in Africa under global warming, Part II: East Africa', *Journal of Climate*, 24, 3718-3733.

Silsbe, G. M., Hecky, R. E., Guildford, S. J. & Mugidde, R., 2006, 'Variability of chlorophyll a and photosynthetic parameters in a nutrient-saturated tropical great lake', *Limnology and Oceanography*, 51, 2052-2063.

Sitoki, L., Gichuki, J., Ezekiel, C., Wanda, F., Mkumbo, O. C. & Marshall, B. E., 2010, 'The environment of Lake Victoria (East Africa): Current status and historical changes', *International Review of Hydrobiology*, 95, 209-223.

Teresa, F. B. & Casatti, L., 2012, 'Influence of forest cover and mesohabitat types on functional and taxonomic diversity of fish communities in Neotropical lowland streams', *Ecology of Freshwater Fish*, 21, 433-442.

Thieme, M. L., Abell, R., Burgess, N., Lehner, M., Dinerstein, E., Olson, D., Tengels, G., Kaudem-Toham, A., Stiassny, M. L. J. S. & Skelton, D., 2013, Freshwater Ecoregions of Africa and Madagascar: A Conservation Assessment, Washington DC, USA: Island Press.

Thrush, S. F., Hewitt, J. E., Norkko, A., Cummings, V. J. & Funnell, G. A., 2003, 'Macrobenthic recovery processes following catastrophic sedimentation on Estuarine sandflats'. *Ecological Applications*, 13, 14433-1455.

Tierney, J. E., Mayes, M. T., Meyer, N., Johnson, C., Swarzenski, P. W., Cohen, A. S. & Russell, J. M., 2010, 'Late-twentieth-century warming in Lake Tanganyika unprecedented since AD 500', *Nature Geoscience*, 3, 422-425.

Troell, M. & Berg, H., 2008, 'Cage fish farming in the tropical Lake Kariba, Zimbabwe: impact and biogeochemical changes in sediment', *Aquaculture Research*, 28, 527-544.

Tumbare, M. J., 2004, 'The Zambezi River: Its threats and opportunities', 7th River Symposium. Brisbane, Australia.

Tumbare, M. J., 2008, 'Managing Lake Kariba sustainably: threats and challenges', *Management of Environmental Quality: An International Journal*, 19, 731-739.

Turner, G., 1995, 'Management, conservation and species changes of exploited fish stocks in Lake Malawi', in T. J. Pitcher, & P. J. B. Hart, eds, The Impact of Species Changes in African Lakes, pp.365-395: Chapman & Hall Fish and Fisheries Series 18. Springer, Netherlands.

Turner, G. F., Seehausen, O., Knight, M. E., Allender, C. J. & Robinson, R. L., 2001, 'How many species of cichlid fishes are there in African lakes?' *Molecular Ecology*, 10, 793-806.

Tweddle, D., 1983, 'Breeding behavior of the mpasa, Opsaridium microlepis (Gunther) (Pisces: Cyprinidae), in Lake Malawi', *Journal of the Limnological Society of Southern Africa*, 9, 23-28.

Tweddle, D., 2001, 'Threatened Fishes of the World: Opsaridium microlepis (Günther, 1864) (Cyprinidae)', *Environmental Biology of Fishes*, 61, 72-72.

Tweddle, D., 2010, 'Overview of the Zambezi River System: Its history, fish fauna, fisheries, and conservation', *Aquatic Ecosystem Health and Management*, 13, 224-240.

Tweddle, D., Cowx, I. G., Peel, R. A. & Weyl, O. L. F., 2015, 'Challenges in fisheries management in the Zambezi, one of the great rivers of Africa', *Fisheries Management and Ecology*, 22, 99-111.

UN-DESA, 2011, World population prospects: The 2010 revision. Volume I: Comprehensive Tables. ST/ESA/SER.A/313. New York: United Nations.

UNDP, 2006, Human Development Report 2006.

UNEP, 1999, Global Environmental outlook (GEO 2000), Nairobi, Kenya: United Nations Environmental Programme.

UN-Water & FAO, 2007, 'Copying with water scarcity. Challenge of the twenty first century'.

UN-Water, 2013, 'Water scarcity Factsheet', www.unwater.org/publications/publications. detail/en/c/204294/

Van Bocxlaer, B., Schulteiß, R., Plisnier, P. D. & Albrecht, C. 2012. Does the decline of gastropods in deep water herald ecosystem change in Lakes Malawi and Tanganyika? *Freshwater Biology*, 57, 1733-1744.

Van Der Knaap, M., Kamitenga, D. M., Many, L. N., Tambwe, A. E. & De Graaf, G. J., 2014a, 'Lake Tanganyika fisheries in post-conflict Democratic Republic of Congo', *Aquatic Ecosystem Health and Management*, 17, 34-40.

Van Der Knaap, M., Katonda, K. I. & De Graaf, G. J., 2014b, 'Lake Tanganyika fisheries frame survey analysis: assessment of the options for management of the fisheries of Lake Tanganyika', *Aquatic Ecosystem Health and Management*, 17, 4-13.

Verburg, P. & Hecky, R. E., 2009a, 'The physics of the warming of Lake Tanganyika by climate change', *Limnology and Oceanography*, 54, 2418-2430.

Verburg, P. & Hecky, R. E., 2009b, 'The physics of the warming of Lake Tanganyika by climate change', *Limnology and Oceanography*, 54, 2418-2430.

Verburg, P., Hecky, R. E. & Kling, H., 2003, 'Ecological Consequences of a Century of Warming in Lake Tanganyika', *Science*, 301, 505-507.

Vincent, W. F., 2010, 'Effects of climate change on Lakes', in G. E. Likens, ed., Encyclopedia of Inland waters: Biogeochemistry of Inland waters, Dordrecht: Elsevier BV, 611-616pp.

Vollmer, E. K., Bootsma, H. A., Hecky, R. E., Patterson, G., Halfman, J. D., Edmond, J. M., Eccles, D. H. & Weiss, R. F., 2005, 'Deep-water warming trend in Lake Malawi, East Africa', *Limnology and Oceanography*, 50, 727-732.

Wenz, P. S., 2001, Environmental Ethics Today, New York: Oxford University Press. Weston, M., 2015, 'Troubled Waters: Why Africa's largest lake is in grave danger', Slate, 27 March, 2015.

Weyl, O. L. F., Ribbink, A. J. & Tweddle, D., 2010, 'Lake Malawi: fishes, fisheries, biodiversity, health and habitat', *Aquatic Ecosystem Health and Management*, 13, 241-254.

William, M. A. J., 2009, 'Human Impact on the Nile Basin: Past, Present, Future', in H. J. Dumont, ed., The Nile: Origin, Environments, Limnology and Human Use. Dordrecht: Springer Science + Business Media B.V.

Witte, F., De Graaf, M., Mkumbo, O. C., El-Moghraby, A. I. & Sibbing, F. A., 2009, 'Fisheries in the Nile System', in H. J. Dumont, ed., The Nile: Origin, Environments, Limnology and Human Use, Doldrecht, The Netherlands: Springer Science + Business Media B.V, pp. 723-747.

Witte, F., Goldschmidt, T., Wanink, J. H., Oijen, M. V., Goudswaard, K., Witte-Maas, E. & Bouton, N., 1992, 'The destruction of an endemic species flock: quantitative data on the decline of the haplochromine cichlids of Lake Victoria', *Environmental Biology of Fishes*, 34, 1-28.

Witte, F., Seehausen, O., Wanink, J. H., Kishe-Machumu, M. A., Rensing, M. & Gold-
    schmidt, T., 2012, 'Cichlid species diversity in naturally and anthropogenically turbid
    habitats of Lake Victoria, East Africa', *Aquatic Sciences*, DOI 10.1007/s00027-012-
    0265-4.
World Bank, 2010, The Zambezi River Basin. A Multi-Sector Investment Opportunities
    Analysis, Volume 1 Summary Report.
Zengeya, T. A. & Marshall, B. E., 2010, 'The inshore fish community of Lake Kariba half
    a century after its creation: what happened to the Upper Zambezi species invasion?',
    African Journal of Aquatic Science, 33, 99-102.

4

River Health Assessment in East and Southern Africa

Lulu Tunu Kaaya, Taurai Bere, Tinotenda Mangadze,
Ismael Aaron Kimirei and Mzime Ndebele-Murisa

Introduction

Rivers provide a range of critical life-support goods and services to both
ecosystems and human communities. Healthy rivers are fundamental components
of biogeochemical cycles, act as water purification systems and provide water
for drinking, agricultural and industrial purposes; fish and other produce for
consumption; buffers against flooding; and recreational services. Widespread
human-induced degradation of freshwater ecosystems associated with changing
habitat structure, water quantity and quality, and biotic interactions has been
reported in the African region (Magadza and Masendu 1986; Gratwicke
1999; Dallas et al. 2010). The increasing trend in the degradation of the
riverine ecosystems in the region are a consequence of rapidly growing human
populations, land use changes, intensified agriculture, increasing urbanization
and industrialization, all of which tend to compromise the natural flow regimes
and water quality which, in turn, influence the provision of goods and services
(Moyo and Phiri 2002; Dallas et al. 2010). Increased efforts are needed to
restore and conserve the natural functions of ecosystems, which would benefit
biodiversity, and society at local, regional, and global levels (Daily et al. 2000;
Hein et al. 2006). Therefore, the need to assess the ecological condition of rivers
and their capacity for continual and sustainable provision of ecosystem services is
undisputable. A healthy river is one that has the ability to maintain its structure
and function, to recover after disturbance, to support local biota (including
human communities), and to maintain key processes, such as sediment transport,

nutrient cycling, assimilation of waste products, and energy exchange. In recognition of the importance of healthy rivers, a comprehensive analysis of river health biological monitoring and assessment in the eastern and southern regions of Africa is discussed in this chapter with a view to providing valuable guidance for the formulation of management policies and strategies that are useful in the decision-making process by relevant management authorities.

River Health Assessment and Monitoring

River health assessment and monitoring are important in ensuring that river systems maintain and retain their ecological processes and functions, and capacity to provide ecosystem services. Conventional physico-chemical assessment and monitoring have been the backbone of water quality monitoring programmes to assess perturbations in aquatic systems (Dallas et al. 2010). However, this conventional approach is limited in that it;

  i.   has a tendency to underestimate and misdiagnose the sources of impairment (Yoder and Rankin 1998; Karr and Chu 1999),

  ii. gives only a water quality measure at the time of sampling without capturing temporal variations (Rocha 1992; Aidar and Sigand 1993),

  iii. may miss potentially toxic compounds,

  iv. is costly, and

  v.   has inadequate sensitivity in analytical measurements of some low concentration pollutants (Dallas et al. 2010).

Besides conventional physico-chemical assessment, another approach in assessment and monitoring of river health is the use of biological components (bioassessment[1] and biomonitoring[2]), which provide a direct measure of ecological integrity by using biota or their responses to environmental changes (Karr 1991; Pan et al. 1996). Bioassessment methods range from the use of sub-organism (cell or tissue) to ecosystem level. Community-level methods are most widely applied by use of single-metric indices (sensitivity or functional groups metrics or biological traits), multi-metric (biotic indices) and multivariate indices (predictive modelling) (Bonada et al. 2006). Bioassessment approaches are based on biological integrity concepts where biological indicators and indices are used. Aquatic organisms are considered to be particularly good indicators of ecological conditions, or degree of water quality impairment in river systems because they integrate and reflect the cumulative effects of the factors impacting an ecosystem over time (Karr and Chu 2000; Bere and Tundisi 2010; Dallas et al. 2010). Biota and their responses allow detection of long-term environmental effects by the ability of biota to reflect conditions or changes prior to the time of sampling. Both aquatic flora (macrophytes and diatoms) and fauna (fish and macroinvertebrates) are used as biological indicators.

The concepts and principles of bioassessment have been embraced in different parts of the world and effective river bioassessment methods have been developed and applied broadly (Wright et al. 1984; Hilsenhoff 1988; Wright 1994; Simpson and Norris 2000). However, in the east and southern Africa region, the South African Scoring System (SASS) of South Africa (Chutter 1998; Dickens and Graham 2002) is the leading bioassessment method which has been widely tested and applied. SASS has been modified and adopted for use in other African countries as the Namibian Scoring System (NASS) in Namibia (Palmer and Taylor 2004), the Okavango Assessment System (OKAS) in Botswana (Dallas 2009), the Zambia Invertebrate Scoring System (ZISS) in Zambia (Lowe et al. 2013), the Tanzania River Scoring System (TARISS) in Tanzania (Kaaya et al. 2015) and the ETHbios in Ethiopia (Aschalew and Otto 2015) (Table 4.1). SASS has also proved to be fairly accurate in evaluating water quality in streams and rivers in Zimbabwe (Ndebele-Murisa 2012; Bere and Nyamupingidza 2014). Recently, Zambia has developed its bioassessment method (Murphy et al. 2015, Table 4.1).

Comparative Use of Biological Indicators

Introduction

A biological indicator or bioindicator is an organism or part of an organism or a community of organisms that can provide information on the quality and ecological status of an ecosystem or a part of an ecosystem. The biological quality and ecological status of aquatic systems can be assessed and monitored using diverse groups of organisms. The efficacy of organisms when used separately has been demonstrated by many studies in the region, e.g. Taylor et al. (2007), Phiri et al. (2007), Bere and Mangadze (2014), Bere et al. (2014), Mangadze et al. (2016) using periphyton, Dickens and Graham (2002), Moyo and Phiri (2002), Ndebele-Murisa (2012), Bere and Nyamupingidza (2014), Kaaya et al. (2015) using benthic macroinvertebrates, Kleynhans et al. (1999) and Kadye (2008) using fish. These groups of organisms have been shown to respond to water chemistry and habitat variability on various scales, making them useful diagnostic and regulatory tools (Karr 1981; Infante et al. 2009). However, each biota has unique properties and unique responses to different types of stress although the appropriateness of the choice of a biological indicator has been questioned (Hilty and Merenlender 2000).

Table 4.1: Summary of bioassessment methods in the east and southern Africa region

Indicator	Biotic Index (Abbreviation)	Taxonomic level	Number of Taxa	Validation	Index range	References
Macro-invertebrates	South African Scoring System (SASS)	Family	99	Yes	1-15	Grahams and Dickens (2002)
	Zambia Invertebrate Scoring System (ZISS)	Family	103	No	1-15	Lowe et al. (2013)
	Namibia(NASS)	Family	90	Yes	1-15	Palmer and Taylor (2004)
	Okavango(OKAS)	Family		No	1-15	Dallas (2009)
	Tanzania River Scoring System (TARISS)	Family	96	Yes	1-15	Kaaya et al. (2015)
	ETHBios (ETHBios)	Family	59	Yes	1-10	Aschalew and Otto (2015)
Diatoms	The South African Diatom Index (SADI)	Species				Harding and Taylor (2011)
Vegetation	Zambia Macrophyte Trophic Ranking System (ZMTR)	Species	225	No		Murphy et al. (2015)

Various studies have identified and grouped several criteria considered important in the selection of biological indicators appropriate for a particular bioassessment (Table 4.2). These criteria provide guidance in selecting bioindicators; however, some of these criteria have received criticism due to lack of clarity and tendency for conflicting signals among them. Specific advantages and disadvantages of using various bioindicators have been further discussed in the next section.

Macroinvertebrates

The most commonly used biological community for river health assessment is the macroinvertebrate community (Walsh 2006). Benthic macroinvertebrates are used for assessing the effects of multiple stressor types such as organic pollution (e.g. Statzner et al. 2001), hydro morphological degradation (Buffagni et al. 2004; Lorenz et al. 2004), acidification (e.g. Sandin et al. 2004) and general stress (e.g. Barbour et al. 1999; Ndebele-Murisa 2012). There are a number of advantages for using benthic aquatic macroinvertebrates in river health assessment programmes including the following (summarized from Hellawell 1986; Metcalfe 1989; Reynoldson and Metcalfe-Smith 1992; Rosenberg and Resh 1993; Metcalfe-Smith 1994; Dallas 2010):

i. Benthic macroinvertebrates are found in most aquatic habitats.

ii. There are a large number of species, and different stresses produce different macroinvertebrate communities.

iii. They have a rapid life cycle often based on seasons.

iv. They are largely non-mobile and thus representative of the location being sampled, which enables effective spatial analyses of pollutant or disturbance effects to be undertaken.

v. They are relatively easy and inexpensive to collect, particularly if qualitative sampling is undertaken.

vi. They are largely visible to the naked eye, and therefore easy to identify.

vii. Their taxonomy is well established, at least to the family level for most groups, with identification keys available.

viii. Their communities are quite heterogeneous, with numerous phyla and trophic levels represented, so there is a high probability that at least some of these organisms will react to a particular change in environmental conditions.

ix. Macroinvertebrates are the primary food source for recreationally and commercially important fish. An impact on macroinvertebrates impacts the food web and designated uses of the water resource.

Table 4.2: Selection criteria for bioindicators as modified from Hilty and Merenlender (2000)

Suggested Criteria	Attributes	
	Clear taxonomy	Taxonomy resolution
	Biology and life history	>30 primary literature articles
	Tolerance level	Studied and validated tolerance levels
Baseline Information	Established correlation to ecosystem changes	Correlation to ecosystem relationships
	Spatial and temporal variabilities	Established trends
Location information	Cosmopolitan vs. Endemic distribution	Global distribution/ Non migratory
	Limited mobility	Small home range size
Niche and Life history characteristics	Early warning	Small body size
	Low variability	Low fluctuations in population
	Easy to find and measure	Easy to find

One disadvantage in the use of macroinvertebrates in bioassessment is their heterogeneous distribution that causes spatio-temporal variability of their communities. Interpretation of biological data may become difficult and less accurate when the degree of patchiness is high.

Periphyton

Periphyton communities are valuable indicators of environmental conditions in aquatic systems. In particular, diatom communities which constitute the major part of the periphyton are composed of a large number of species with various ecological tolerances and preferences, thus, constituting a well-adapted biological model for environmental monitoring (Bere and Tundisi 2011; Bere and Mangadze 2014). De la Rey et al. (2004) and Harding et al. (2005) list several advantages of using diatoms in river health assessment:

i.   They have rapid reproduction rates and very short life cycles and therefore can be early warning systems for environmental degradation.

ii. As primary producers, periphyton act as important foundation of food webs in river ecosystems and therefore bear cascading impacts up the food chain.

iii. The large number of taxa provides redundancies of information and important internal checks in datasets, increasing the confidence of environmental inferences.

iv. Diatom frustules have a lasting permanence in sediments, such that sediment cores provide details of changes in the quality of the overlying water for as far back as one is able to search.

v.   Diatoms respond rapidly to eutrophication (a major challenge in some parts of the Zambezi Basin, especially near urban areas) because by their photoautotrophic natures, diatom growth is directly affected by changes in prevailing nutrient concentrations.

vi. They are easy to collect, prepare for observation, and to store (small sample volumes, no desiccation risk) for reference purposes.

The use of periphyton, however, is limited by the difficulty in accurate taxonomical identification particularly to species level which is important for effective use of diatoms in bioassessment. It also takes longer time for results of identification to species level to be available.

Fish

Fish communities respond significantly and predictably to almost all kinds of anthropogenic disturbances, including eutrophication, chemical pollution, flow regulation, physical habitat alteration and fragmentation, human exploitation and introduced species (Kleynhans et al. 1999; Ormerod 2003; Kadye 2008). Below are some of the advantages of using fish as biomonitoring tools:

i.   Fish are good indicators of long-term (several years) effects and broad habitat conditions because they are relatively long-lived and mobile (Karr et al. 1986).

ii. Fish are at the top of the aquatic food web and are consumed by humans, making them important for assessing contamination.

iii. They are relatively easy to collect and identify to the species level.

iv. Environmental requirements, life history information and distribution are well known for most species.

v.   Monitoring fish provides direct evaluation of "fish propagation", which emphasizes the importance of fish to fishermen.

Despite the advantages highlighted above, using fish communities in biological monitoring has some disadvantages in that due to mobility and migration, it is difficult to pinpoint a pollutant as the cause of abnormalities in individuals, population, or communities. Given the threats faced by many rivers in the region, there is a need for appropriate and scientifically validated bioassessment methods based on diverse bioindicators. Each biological indicator has some limitations and hence the use of multiple indicators is now considered a key tool for river health assessment. The use of multiple indicators increases the probability of detecting change (multiple lines of evidence) (Johnson 2006). This knowledge of how organisms respond to different types of stress can and should be used to design robust and cost-effective monitoring programmes.

Case Study 1: Comparative Use of Diatoms, Fish and Macroinvertebrates in Biomonitoring in Zimbabwe

Developing appropriate quantitative tools for river health assessment is a pressing need in Zimbabwe. River health assessment has traditionally been through the physico-chemical approach, with biological monitoring being rare and incipient in Zimbabwe. This study explores the potential for biological monitoring in Zimbabwe and compares the responses of diatoms, macroinvertebrates, and fish to human impacts in the Manyame catchment. The study aims to promote the use of biomonitoring by relevant government organs and non-governmental organisations responsible for conservation and management of river systems.

Study Area

A total of 44 sampling stations was selected and sampled in four land-use settings: commercial agriculture (n = 15); communal agriculture (n = 14); and mining (n = 6); and urban (n = 9) (Figure 4.1); which represent a gradient in river health status. For details of land uses and other characteristics in this catchment, please see Chapter 2 of this volume. Commercial agricultural areas were expected to be relatively clean compared to communal, mining and urban areas. These areas were characterized by mature deciduous riparian forest strips which acted as riparian buffers thus protecting water resources from nonpoint sources of pollution and providing bank stability and aquatic habitats. Communal agricultural areas suffer from a combination of poor agricultural practices (stream bank cultivation, overgrazing, soil erosions) and high human population densities leading to detrimental effects on river health. The mining areas showed signs of ecosystem degradation and poor river health due to the increase in the number of illegal gold panners that use methods, which are destructive to the natural environment. Streams draining urban areas received pollutants from various points and diffuse sources and their habitats have been greatly altered resulting in stream health deterioration, eutrophication, organic and metal pollution among other threats. Two samplings were carried out in April (end of rainy season) and September (during the dry season) 2013 to capture the two flow extremes typical of the study area (Meteorological Services Department of Zimbabwe, 1965-2014).

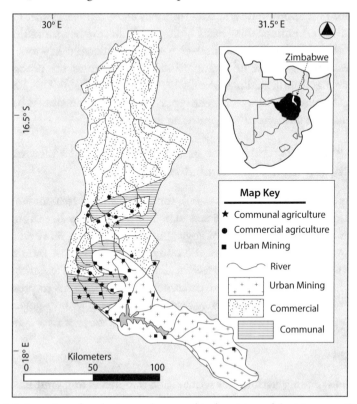

Figure 4.1: Location of study sites and geographical position of the Manyame Catchment in Zimbabwe, Africa

Biota Sampling and Analysis

At each site, epilithic diatoms were sampled from stones, sorted in the lab, counted and identified to species level following Taylor et al. (2005). The trophic diatom index (TDI), described by Kelly and Whitton (1995), was calculated and used to assess water quality at the different sampling sites. Macroinvertebrate samples were collected following the South African Scoring System version 5 protocol (SASS5) (Dickens and Graham 2002). SASS5 scores and average score per taxon (ASPT) were calculated following (Dickens and Graham 2002). Fish were collected from riffle, run, and pool habitats as described by Moulton et al. (2002). Fish samples were counted and identified to species level in the field using standard taxonomic guides (Skelton 2001; Marshall 2011). The Fish Assemblage Index (FAII) was used to determine the status of the fish assemblages' integrity in relation to disturbances in lotic ecosystems following Kleynhans et al. (1999). Canonical correspondence analysis (CCA) was performed per each assemblage group to relate community structure to simultaneous effects of predictor variables (Ter Braak and Verdonschot 1995) and to determine whether diatom, macroinvertebrate and fish communities responded to the same environmental gradients using CANOCO version 5 (Ter Braak and Šmilauer 2002).

Water Quality Sampling and Analysis

Along with biota sampling at each station, the following variables were measured: electrical conductivity (EC), dissolved oxygen (DO), total dissolved solids (TDS), salinity, conductivity and temperature, magnesium (Mg), calcium (Ca), nickel (Ni), potassium (K+), sodium (Na+), lead (Pb2+), zinc (Zn2+), iron (Fe2+), chromium (Cr3+), cadmium (Cd2+), manganese (Mn2+), copper (Cu2+), total hardness (TH), total phosphate (TP), soluble reactive phosphate (SRP), total nitrogen (TN) and chemical oxygen demand (COD) (APHA 1988). A two-way analysis of variance (Two-Way ANOVA) with Tukey's post hoc HSD tests was used to compare means of physical and chemical variables among sampling stations and the compounded effects of season.

Results and Discussion

The values of the physiochemical variables recorded in the Manyame catchment during this study are summarized in Table 4.3. The water quality generally tended to deteriorate in urban sampling stations with some of the parameters. No significant differences were observed in mean environmental variables between the two sampling periods (t-test, p>0.05). DO was significantly lower in communal and urban areas as compared to agricultural and mining areas (ANOVA, p<0.05). TDS and conductivity were significantly low in commercial farming areas compared to the rest of the sites (ANOVA, p<0.05). TP, SRP, TN and COD were significantly higher in urban sampling stations compared to the rest of the sampling stations (ANOVA, p < 0.05). Ni^{2+} was significantly higher in mining areas compared to the rest of the sampling sites (ANOVA, p<0.05). K^+, Na^+, CA^{2+} were significantly high in urban and mining sites compared to the rest of the sites (ANOVA, p<0.05).

Table 4.3: The mean, standard deviations for physico-chemical characteristics recorded in streams of the Manyame Catchment, Zimbabwe analysed per riparian-scale land use (Superscript letters indicate values that significantly differ with others in the same row (Tukey's HSD, p<0.05). * indicates that they are no standard limits for the parameter)

Parameter	Categories of sampling stations			
	Commercial Farming	Communal Farming	Great Dyke Mining	Urban
Temperature (°C)	20.5±1.65	22.5±1.95	20.24±2.15	20.4±1.98
DO (mgl⁻¹)	5.38±1.49a	3.18±1.66[b]	4.52±1.56ab	3.1±1.31[b]
Conductivity (µScm⁻¹)	287.9±166.97[a]	365.2± 157.83[b]	393.76±169.13[b]	382.6±147.9[b]
TDS (mgl⁻¹)	200.6±106.96[a]	253.4±115.55[b]	278.9±127.04[b]	273.1±111.09[b]
Salinity (ppt)	0.14±0.08a	0.19±0.08[a]	0.11±0.04[b]	0.20±0.05[a]
TN (mgl⁻¹)	2.31±1.17[a]	3.36±1.71[b]	2.5±1.06[a]	7.8±3.6[c]
TP (mgl⁻¹)	0.09±0.06[a]	0.13±0.08[a]	0.06±0.05[a]	0.4±0.33[b]
SRP (mgl⁻¹)	0.01±0.01[a]	0.09±0.03[b]	0.06±0.04[b]	0.16±0.05[c]
COD (mgl⁻¹)	105.2±71.43	104.2±89.46	112.3±77.5	127.3±78.2
Mg²⁺(mgl⁻¹)	13.5±4.12[a]	15.80±3.29[a]	23.6±3.25[b]	17.1±3.9[ab]
Ni²⁺ (mgl⁻¹)	0.04±0.03[a]	0.02±0.01[a]	0.09±0.02[b]	0.03±0.01[a]
Total hardness (mgl⁻¹)	83.86±34.92	105.98±34.46	89.67±38.75	95.3±36.1
Ca²⁺(mgl⁻¹)	11.67±2.3[a]	16.38±2.41[b]	20.38±2.62[c]	17.1±2.22[b]
K⁺ (mgl⁻¹)	2.07±1.0[a]	1.59±1.33[a]	6.2±1.73[b]	4.1±1.92[b]
Na⁺ (mgl⁻¹)	11.18±6.68[a]	11.09±6.31[a]	21.39±7.1[b]	24.3±6.2[b]

The TDI displayed an oligotrophic state in commercial agricultural sampling stations (32±13.8), communal agricultural sampling stations (34±12) and mining sampling stations (33.5± 8.15, Table 4.4). The index score for urban sites (51.1±14.7) suggest a reduction in water quality as the sampling stations are in a fairly eutrophic state. Generally, all the sampling stations had SASS scores below 100, with the highest mean score of 51.6±12.1 recorded at communal agricultural sampling stations and the lowest score of 37.4 ±10.1 at urban sampling stations indicating a general deterioration in water quality among all sampling station categories. Based on the ASPT index values, ecological health at all the sites ranged from fair to poor with relatively low habitat diversity. Low SASS (37.4±10.1) and ASPT (4.2±1.2) index scores were recorded at urban sampling stations compared to the rest of the sites, indicating a very poor ecological state with signs of major deterioration in water quality and also low habitat diversity. Relative FAII scores of 62.2, 67.6 63.0 and 61.0 per cent were recorded for commercial, communal, urban and mining areas respectively indicating moderately modified conditions of all the four site categories (Table 4.4).

Table 4.4: Mantel test amongst biological data (diatoms, macroinvertebrates and fish) and physical-chemical parameters in the Manyame catchment, Zimbabwe, in 2013 (rm = Mantel correlation coefficient; p = p-values; *, $p > 0.05$, not significant)

Physical and Chemical Varibale	Biological Element					
	Diatom		Macroinvertebrates		Fish	
	rm	p	rm	p	rm	p
Eutrophication	0.28	0.02	0.16	0.03	-0.08	0.88*
Metal pollution	0.45	<0.01	0.06	0.16*	0.01	0.44*
Organic Pollution	0.42	<0.01	0.10	0.05	0.03	0.32*
Other variables	0.44	<0.01	0.10	0.06*	0.003	0.45*

The CCAs performed to relate diatom, macroinvertebrate and fish community structure to simultaneous effects of predictor variables explained more than 50 percent of the biotic species variance in all cases and in most cases managed to roughly separate sampling stations into 4 groups; commercial agriculture, communal agriculture, urban and mining gradient (human induced increase in nutrients, organic and metal pollution and decrease in DO). This indicates the ability of these bioindicators to detect changes induced by human disturbance (Figure 4.2).

Diatom assemblages were shown to be more related to eutrophication, metal pollution and organic pollution. Correlation amongst macroinvertebrates and eutrophication as well as organic pollution appeared weaker, but still significant; whilst for metal pollution there was no significant correlation for any biota. On the other hand, there was no significant correlation between fish assemblages and all measured environmental variables (Table 4.4).

Conclusions: Manyame Case Study

Organism response to stress varied among organism groups and with different types of stress. This implies that choice of the 'best' organism group and/or metric is crucial for detecting (or not detecting) human-induced change in stream integrity. Based on biotic indices and results of multivariate analyses, diatoms and macroinvertebrate were found to be much better than fish in indicating degree of pollution as well as in discriminating less impacted sampling stations from highly impacted sampling stations of the study area. In addition, African

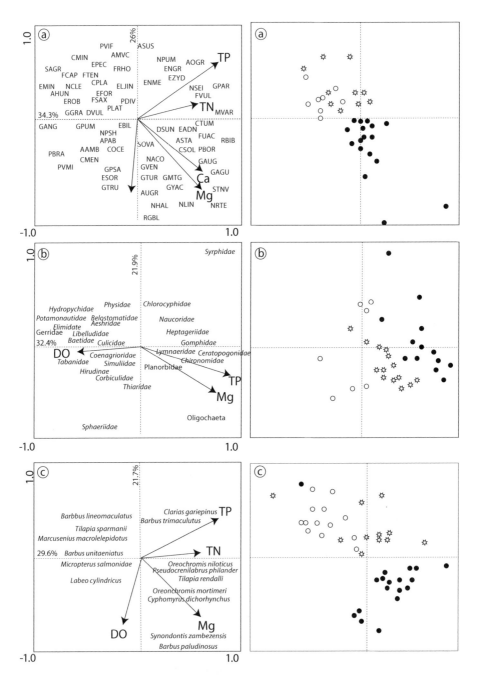

Figure 4.2: Canonical correspondence analysis (CCA) in 44 sampling stations: (a) diatoms (b) macroinvertebrates (c) fish and the corresponding land use sampling stations with circles = Commercial agricultural (Group 1); down-triangle = great dyke, mining (Group 2); cross-hatch = communal agricultural (Group 3) and diamonds = urban (Group 4)

highland low order streams have low fish diversity, making the use of diatoms and macroinvertebrates more pertinent. Benthic diatoms showed strong association with variables related to eutrophication while macroinvertebrates were more associated with organic pollution. Thus, diatoms and macroinvertebrates are complementary to each other in monitoring programs.

Case Study 2: Bioassessment of Rivers Using Aquatic Macroinvertebrates in Tanzania

The use of biota for assessment and monitoring of river systems is still a new approach in Tanzania which started in 2006 in the Pangani River Basin (PBWO/IUCN 2007). The Pangani River Health Assessment (RHA) using fish, macroinvertebrates and vegetation was conducted by adopting bioassessment methods from South Africa, e.g. SASS5 was used for macroinvertebrates. River health assessment expanded to other basins and sub basins in the country including Great River Ruaha basin (Mwakalila and Masolwa 2012), Wami and Ruvu basins (GLOW_FIU 2014) and Sigi basin. However, lack of appropriate validated bioassessment methods and biomonitoring programmes became a major constraint to successful river health assessment in the country. In addition, lack of several bioassessment tools such as homogenous river classifications, standard bioassessment methods and reference conditions for robust interpretation of data and synthesis of information were hindrances to effective biological RHA.

In 2015, a milestone development in bioassessment for Tanzanian rivers was achieved when a river classification approach and a bioassessment method were developed and validated in selected river basins. Rivers were classified into homogenous regions following the established two-level hierarchical classification framework (Figure 4.3) and validated using aquatic macroinvertebrates.

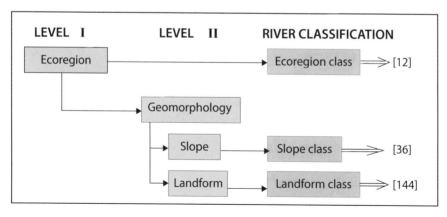

Figure 4.3 A hierarchical spatial framework for classification of rivers in Tanzania

Classification of river systems as a component of water resources is stipulated as a requirement in the Tanzania National Water Policy of 2002. The classification framework was validated in three river basins (Pangani, Rufiji and Wami-Ruvu) (Kaaya 2015). River classification is an essential tool for biological river health assessment hence this is considered a significant development for river bioassessment in Tanzania.

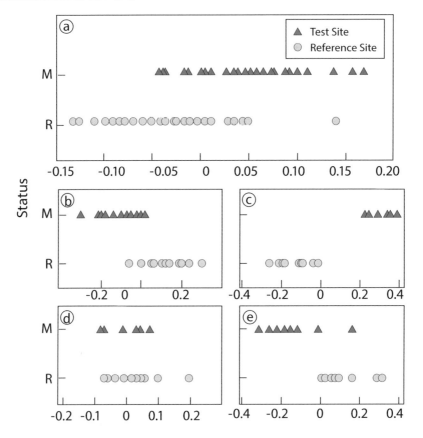

Figure 4.4: Canonical analysis of principal coordinates (CAP) ordination, showing the position of the new macroinvertebrate taxa in a gradient of macroinvertebrate sensitivity relative to sensitivity groups in the Pangani, Wami-Ruvu and Rufiji basins, Tanzania, in 2010-2012

A macroinvertebrate-based bioassessment method for Tanzanian rivers, the Tanzania River Scoring System (TARISS) was also developed and validated in 2015 (Kaaya et al. 2015). Development of TARISS was also in line with objectives of the Ministry of Water in providing water of acceptable quality. Supply of water of acceptable quality can partly be achieved through development and implementation of practical, cost-effective water quality and pollution control assessment and monitoring programmes (URT 2002) such as bioassessment.

TARISS was developed from the South African Scoring System (SASS) and validated for use in Tanzanian rivers. Differences in climate, geology, longitude and latitude between Tanzania and South Africa may lead to differences in the riverine physical and chemical characteristics, and macroinvertebrate assemblages and their sensitivity levels in relation to disturbance and general ecosystem impairment (Kaaya et al. 2015). Such variation could affect the functioning and reliability of the SASS method when directly applied in Tanzanian rivers without modification. Through validation, TARISS proved to be reliable in distinguishing reference from test sites, based on both macroinvertebrate assemblages and TARISS metrics. In the validation, TARISS was evaluated along the human disturbance gradient using macroinvertebrate assemblages to test its ability to distinguish sites (Figure 4.4). TARISS is part of the initial steps in the development of bioassessment tools including the national or regional biotic index for assessment of rivers.

Furthermore, river bioassessment using TARISS have been used in assessing health conditions at small basin scales within the country including Lake Tanganyika Basin, Lake Nyasa Basin and Umba Basin in 2014 and within the internal drainage Basin including the Tarangire River catchment 2015 (Plate 4.1). Riverine bioassessment in Tanzania has currently started penetrating involvement of local communities and water user association officers. Currently, groups of water user association leaders in the Mkoji catchment, Rufiji Basin, are being trained for biomonitoring using a simplified bioassessment method (Plate 4.2).

Plate 4.1: River health bioassessment using aquatic macroinvertebrates in Tarangire, Tanganyika, Umba and Nyasa basins, Tanzania, 2014

Plate 4.2: Practical training of water user association leaders on bioassessment using aquatic macroinvertebrates in Mkoji Catchment, Rufiji Basin, Tanzania, 2016

It is important to understand that apart from classification of rivers and development of the TARISS bioassessment method, there are other important bioassessment aspects including characterization of reference conditions for interpretation of results. A step flow diagram showing the steps from identification of homogenous regions to characterization of reference conditions is shown in Figure 4.5. This is important for attaining effective bioassessment.

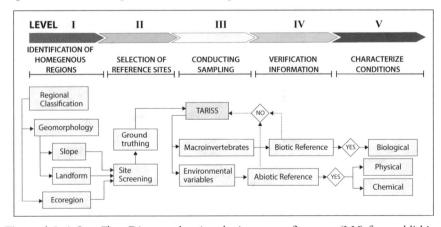

Figure 4.5: A Step Flow Diagram showing the important five steps (I-V) for establishing reference conditions for rivers, Tanzania

The Influence of Seasonality on Bioassessment: A Case Study of Wami-Ruvu, Pangani and Udzungwa Catchments, Tanzania

Lotic systems in tropical regions exhibit seasonal variation in hydrology (Lewis 2008) due to climatic seasonality in the form of rainfall which occurs in alternating wet and dry periods (Jacobsen et al. 2008). Hydrological variation is usually associated with seasonal variation in flow, depth and velocity, stream width (e.g. Minshall et al. 1985), water chemistry, sediment transport, allochthonous inputs and metabolic rates (e.g. Lewis 2008). Temporal variation in these physical and chemical variables may result in natural temporal variation in macroinvertebrate assemblages. The inherent temporal variation in macroinvertebrate assemblages may influence bioassessment indices or metrics and their interpretation. For instance, comparison of test samples collected in a particular sampling period with reference conditions derived from another sampling period is likely to result in biased conclusions regarding ecological status of the test site (e.g. Linke et al. 1999; Reece et al. 2001). A study by Linke et al. (1999) in Ontario, Canada showed that bioassessment using taxon richness and Family Biotic Index metrics in winter resulted in a higher water quality status than in summer, while a predictive model developed by Reece et al. (2001) in British Columbia using macroinvertebrate reference data from autumn could not be used to accurately predict macroinvertebrates at test sites in other seasons because of seasonal variation.

The primary objective of bioassessment is to detect the degree of impairment on a test site by comparing it to a reference condition. It is important to ensure the accuracy and reliability of the reference condition by understanding the effects of seasonal variability (Dallas 2004). Temporal variability has been controlled or reduced either by collecting samples within a short period of time (i.e. in established index periods: Barbour et al. 1999) or by combining multiple season samples of a site (Wright 2000). Combined season datasets have been shown to increase the prediction accuracy of faunal composition of test sites (Furse et al. 1984). However, the increased time and costs of data collection in multiple seasons creates a negative implication in bioassessment. Prior to the application of bioassessment for river health assessment, it is important to examine and characterize temporal variability of biotic assemblages, indices and metrics in order to avoid incorrect inferences. This study objectives are to: 1) to examine temporal variation in TARISS taxa, macroinvertebrate assemblages, number of taxa, TARISS scores and ASPT; and 2) to test the hypothesis that combined seasons reference data increases the accuracy of distinguishing test sites from reference sites.

Methods and Study Area

An area of 58,100 km² in Tanzania is covered by freshwater distributed in watersheds of four lake basins (Victoria, Tanganyika, Nyasa and Rukwa), four river basins (Pangani, Rufiji, Ruvuma, Wami-Ruvu) and one internal drainage basin (URT 2011). This study covered three basins, namely Pangani, Wami-Ruvu and Rufiji (Figure 4.6). The Pangani basin is characterized by Pangani and Tana ecoregions while Wami-Ruvu and Rufiji basins occur in the coastal eastern Africa ecoregion. A total of 101 sites including reference and test sites were sampled for macroinvertebrates during the dry and wet periods.

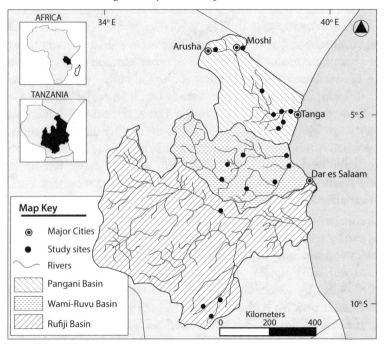

Figure 4.6: Location of study sites and geographical position of the study area in Tanzania

Macroinvertebrate Sampling

Macroinvertebrates were sampled in wet (long rains and short rains) and dry seasons using the TARISS (Kaaya et al. 2015). Data from three river types, namely Pangani highland uplands (PHU), Central Eastern Africa uplands (CEAU) and Central Eastern Africa Lowlands (CEAL) were analysed in combined biotopes data set (stone, vegetation, gravel sand mud) and stone biotopes data set. The three river types were selected because they had adequate number of sites sampled across seasons. Vegetation and GSM biotopes were not singly analysed because they occurred in fewer sites compared to stones; hence stone biotope and a set of combined biotopes were used.

Data Analysis

Individual Taxa

Frequency of occurrence of individual TARISS taxa was calculated separately for the long rains, short rains and dry sampling periods by counting the number of times a taxon occurred among sampling occasions divided by the total number of sampling occasions and expressed as a percentage. Taxa from different biotopes were combined (TARISS total taxa at a site).

Macroinvertebrate Assemblages

ANOSIM was used to test the hypothesis that there was no difference in community patterns among the sampling periods within each river type (Clarke and Gorley 2006). Cluster analysis was used to visualize grouping of macroinvertebrate assemblages between wet and dry sampling periods in CEAL because ANOSIM showed significant differences in CEAL only (Clarke and Gorley 2006). SIMPER analyses were used to identify taxa responsible for within-group similarities and between group dissimilarities (Clarke and Gorley 2006). Principle coordinate ordinations were used to show clustering of sites between sampling periods and for exploration of macroinvertebrate taxa contributing to the clustering (Anderson et al. 2008).

Number of Taxa, TARISS Scores and ASPT

Number of taxa, TARISS scores and ASPT were compared among the sampling periods using one-way ANOVA after passing the Kolmogorov-Smirnov and Liliefurs test for normality in Statistica at 10 intervals. Number of taxa, TARISS scores and ASPT from single sampling period datasets (i.e. dry, long rains and short rains), were compared with number of taxa, TARISS scores and ASPT calculated from a combined sampling period dataset using one-way ANOVA. Percentage contribution of number of taxa, TARISS scores and ASPT of each sampling period to combined sampling period values were calculated. TARISS metrics from single sampling periods at test sites were compared with metrics from reference data collected in a single period and combined periods for PHU, CEAU and CEAL to compare the two reference datasets.

Results

Frequency of Occurrence of TARISS Taxa Among Sampling Periods

Frequencies of occurrence of individual TARISS taxa in each sampling period for PHU, CEAU and CEAL were calculated, giving relative frequency of occurrence (%) of TARISS taxon in the long rains (L), dry (D) and short rains (S) in the Pangani Highland Uplands (PHU), Central Eastern Africa Uplands (CEAU) and Central Eastern Africa Lowlands (CEAL) river types.

Many taxa occurred in all sampling periods for all river types while a few taxa were present or absent, or occurred more or less frequently in certain sampling periods. In PHU, CEAU and CEAL, 10, 7 and 13 taxa respectively showed temporal preferences by occurring in higher frequencies in a particular sampling period. In PHU and CEAU, the majority of taxa that indicated temporal preferences occurred at lower frequencies in the long rains than in the short rains and dry periods. In CEAL, taxa associated with stone and fast flowing biotopes such as the Perlidae, Heptageniidae, Leptophlebiidae, Prosopistomatidae, Psephenidae and Athericidae occurred at higher frequencies during the wet than dry period. On the contrary, vegetation-associated taxa such as the Coenogrionidae and Naucoridae and slow flow, stream edges and pool-associated taxa such as the Gerridae, Veliidae, Chironomidae and Dytiscidae occurred in higher frequencies during dry period than the wet period. Prosopistomatidae did not occur in the dry period in CEAL. Only five taxa showed temporal preferences in more than one river type. The Coenogrionidae, Velidae and Dytiscidae occurred in higher frequencies in the dry period, the Hepategeniidae occurred in higher frequencies in wet periods and the Chironomidae occurred in higher frequencies in the dry period in CEAL, and in dry and short periods in PHU and CEAU. Potamonautidae, Chironomidae, Gomphidae, Leptophlebiidae, Coenogrionidae, Baetidae > 2sp, Elmidae, Tipulidae and Veliidae occurred in higher frequencies among the sampling periods than other taxa in all river types.

Temporal Variation in Macroinvertebrate Assemblages

Macroinvertebrate assemblages did not show any temporal variation among the sampling periods in PHU (global $R = 0.009$, $p = 0.401$ and global $R = 0.04$, $p = 0.204$) or CEAU (global $R = 0.015$, $p = 0.326$ and global $R = 0.104$, $p = 0.051$) in both combined biotopes and stone biotope respectively. In contrast, in the CEAL, macroinvertebrate assemblages showed significant temporal differences between the wet and dry periods in both combined and stone biotopes (global $R = 0.201$, $p = 0.005$ and global $R = 0.034$, $p = 0.001$ respectively). Cluster analysis in CEAL illustrates grouping patterns of macroinvertebrate assemblages in combined biotopes and stone biotope (Figure 4.7). Analyses for similarity levels within the wet and dry periods were performed using SIMPER analysis by the Bray-Curtis similarity with a 90 per cent cut off for low contributions. In combined biotopes analysis, within-wet period similarity was 50.74 per cent contributed by Potamonautidae, Leptophlebiidae, Gomphidae, Elmidae, Heptageniidae and Prosopistomatidae, and the within-dry period similarity was 53.38 per cent contributed by Gomphidae, Potamonautidae, Veliidae, Ephemerythidae, Coenogrionidae and Baetidae >2sp. Analysis of stone biotope dataset revealed stronger differences between the wet and dry periods with the within-wet period similarity of 62.76 per cent contributed by Potamonautidae,

Leptophlebiidae, Prosopistomatidae, Psephenidae, Heptageniidae, Perlidae and Baetidae > 2sp; and the within-dry period similarity of 66.15 per cent contributed by Potamonautidae, Ephemerythidae, Gomphidae, Baetidae >2sp Elmidae and Tabanidae, as shown in Figure 4.7.

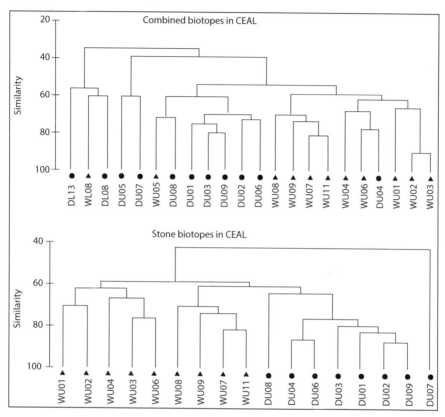

Figure 4.7: Dendrogram shows the clustering of macroinvertebrate assemblages between wet and dry sampling periods in combined biotopes and stone biotope in CEAL (Wet = ● and Dry = ▲)

Temporal Variation in TARISS Metrics

In the combined biotopes, the only significant difference in TARISS metrics amongst sampling periods was for ASPT in PHU ($F = 3.504$, $p = 0.0407$) and CEAL ($t = 2.896$, $p = 0.0084$, $DF = 22$). In the stone biotope, significant differences between wet and dry sampling periods were found in CEAL only for number of taxa ($t = 2.565$, p = 0.0215, DF = 15), TARISS scores ($t = 3.971$, $p = 0.0012$, DF = 15) and ASPT (U'= 60, $p = 0.0206$).

Table 4.5: One-way analysis of variance (ANOVA) of TARISS metrics between combined periods, long rains, dry, short rains references and long rains, dry, short rains test sites (LT = long rains test, ST = short rains test, DT = dry test)

TARISS metrics	PHU	q	p	CEAL	q	p
	CR vs LT	8.229	<0.001	CR vs LT	8.68	<0.001
	CR vs DT	9.842	<0.001	CR vs DT	6.783	<0.001
	CR vs ST	7.94	<0.001			
Number of Taxa	LR vs LT	0.207	>0.05	LR vs LT	5.274	<0.001
	LR vs T	1.856	>0.05	LR vs T	3.085	>0.05
	LR vs ST	0.7575	>0.05			
	DR vs LT	3.125	>0.05	DR vs LT	5.94	<0.001
	DR vs DT	4.774	<0.05	DR vs DT	3.814	>0.05
	DR vs ST	3.307	>0.05			
	SR vs LT	1.729	>0.05			
	SR vs DT	3.378	>0.05			
	SR vs ST	2.088	>0.05			
TARISS Score	CR vs LT	9.126	<0.001	CR vs LT	9.457	<0.001
	CR vs DT	10.548	<0.001	CR vs DT	8.597	<0.001
	CR vs ST	8.631	<0.001			
	LR vs LT	1.547	>0.05	LR vs LT	6.903	<0.001
	LR vs DT	2.962	>0.05	LR vs DT	5.799	<0.001
	LR vs ST	2.008	>0.05			
	DR vs LT	4.696	<0.05	DR vs LT	6.303	<0.001
	DR vs DT	6.118	<0.05	DR vs DT	5.142	<0.001
	DR vs ST	4.76	<0.05			
	SR vs LT	2.125	>0.05			
	SR vs DT	3.548	>0.05			
	SR vs ST	2.514	>0.05			
	CR vs LT	2.817	>0.05	CR vs LT	5.476	<0.001
	CR vs DT	4.337	<0.05	CR vs DT	7.356	<0.001
	CR vs ST	2.581	>0.05			
ASPT	LR vs LT	3.583	>0.05	LR vs LT	7.65	<0.001
	LR vs DT	5.103	<0.01	LR vs DT	9.669	<0.001
	LR vs ST	3.25	>0.05			
	DR vs LT	3.77	>0.05	DR vs LT	4.19	<0.05
	DR vs DT	5.289	<0.01	DR vs DT	5.858	<0.001
	DR vs ST	3.413	>0.05			
	SR vs LT	1.503	>0.05			
	SR vs DT	3.023	>0.05			
	SR vs ST	1.433	>0.05			

Comparisons of TARISS metrics between single sampling periods (i.e. dry, long rains and short rains) and combined sampling periods using combined biotope data sets for PHU showed significant differences between single and combined sampling periods in the number of taxa ($p = 0.0008$, $F = 6.954$) and TARISS scores ($p = 0.0009$, $F = 6.917$) where significant differences were between the combined period and long rains and short rains. No significant differences were found between metrics for the combined period and dry period's metrics. In CEAU, significant differences occurred only in the number of taxa ($p = 0.0016$, $F = 5.853$) between combined periods and dry and long rains, while in CEAL there were no significant differences between combined periods and dry or wet periods in either number of taxa ($p = 0.1318$, $F = 2.145$) nor TARISS score ($p = 0.1145$, $F = 2.303$). ASPT only varied between wet and dry periods in CEAL ($p = 0.01$, $F = 5.249$). Combined sampling period's data resulted in more significant differences between reference and test sites than did single sampling period's data for all seasons. Significant differences were in the number of taxa and TARISS score in PHU, and in the number of taxa, TARISS score and ASPT in CEAL (Table 4.5), which gives a summary of statistical comparison between reference and test metrics.

One-way analysis of variance (ANOVA) of TARISS metrics between combined periods, long rains, dry, short rains references and long rains, dry, short rains test sites. The Turkey-Kramer multiple comparison test (q) determined significant differences between groups in PHU and CEAL. Shaded cells indicate comparisons with significant differences. (CR = Combined periods reference, LR = long rains reference, SR = short rains reference, DR = dry reference.

Discussion

Tanzanian Case Study

In this case study, major temporal variation in the frequency of occurrence of individual TARISS taxa among different sampling periods was limited to a few taxa and varied among river types. Higher frequencies of lithophilic taxa in the wet period than the dry period may be due to influences of changes in flow regime or availability of stony biotopes as well as their stability. The Perlidae, Heptageniidae, Leptophlebiidae, Prosopistomatidae, Psephenidae, Athericidae and Hydropsychidae are all known to occur in stony or rocky substrate in moderately to fast-flowing streams. In the wet period, flow depth and velocity increase, hence availability of riffles increased, and supported the occurrence and dominance of these taxa. For example, the Prosopistomatidae, which showed a marked occurrence in the wet period only in CEAL, is known to be lithophilic and rheophilic because its distribution is driven primarily by substratum and current features (Schletterer and Fureder 2009). In contrast, higher frequencies of the Coenogrionidae, Gerridae, Naucoridae, Veliidae, Dytiscidae and

Chironomidae that inhabit vegetation, stream edges and pools may also be associated with reduction in flow which supports availability of vegetation, and pool-associated habitats. In addition, allochthonous materials such as leaves and twigs accumulated during the dry period as a result of slow transport of materials due to low flow velocity may have provided food and surfaces for attachment of these taxa (Mathooko and Mavuti 1992).

In East African tropical region, flow regime and habitat stability as functions of seasonality, have a primary influence on macroinvertebrate life cycles and population dynamics (Jacobsen et al. 2008); temperature is less important because of minimal variability in mean annual temperatures (Griffiths 1972; Mwamende 2009). The seasonal variability in the frequency of occurrence of TARISS taxa in this study is important in biological assessment of rivers and streams because taxa with higher frequencies in the wet period tend to be the more sensitive taxa (sensitivity weightings of 10-5) while the dry period is dominated by more tolerant taxa (sensitivity weightings of 2-7). This means the wet period samples will reflect a higher ecological category than the dry period samples at the same site leading to two different ecosystem status values. Understanding and considering the temporal variation in the three regions of this study assist in the interpretation of bioassessment data, and allow for temporal variability to be accounted for in bioassessment.

Macroinvertebrate assemblages showed temporal variation only in CEAL and did not vary significantly in PHU or CEAU. These results can be explained by the frequency of occurrence results, where more taxa in CEAL showed temporal variation than in PHU and CEAU. Furthermore, CEAL was the only river type that included sites from unimodal rainfall regions which have one rainy sampling period and one dry period. Truly seasonal conditions in the tropics occur in rivers within the moonsonal regions where there are marked wet and dry periods (Jacobsenet al. 2008). Unimodal rainfall patterns generate a stronger difference between the wet and dry periods than bimodal rainfall patterns where the difference between the wet and dry period is reduced by the short rains. Marked seasonal conditions transform into marked seasonal changes in flow regime, sediment transport, allochthonous materials supply, habitat availability and stability which in the end define macroinvertebrate life histories and distribution. As a result, macroinvertebrate taxon composition changes less annually in bimodal regions than in unimodal regions. This is relevant to the issue of appropriate TARISS sampling period. The choice of when TARISS samples are collected should be determined based on the rainfall pattern at a given region and the knowledge of differences in macroinvertebrate assemblages brought by the wet and dry period cycles should be considered in the interpretation of data.

Examination of macroinvertebrate   assemblages in the CEAL stone biotope resulted in a clearer differentiation between the rainy and dry periods than when combined biotopes were used, possibly because the stone biotope is more sensitive to flow regime changes than the other biotopes are (Dallas 2002). Dallas (2002) found stronger differences in macroinvertebrate assemblages in stone-associated biotopes between spring and autumn periods in South Africa than in combined biotopes. In this study, when stone biotope sample was used, the taxa contributing to within group similarity in the wet period included more taxa associated with stone, fast-flowing biotope than in combined biotopes. This resulted in increased differences between the wet and dry sampling periods compared to when combined biotopes was used whereby CEAL substrata were characterized by gravel and small stones including pebbles and cobbles. Taxa such as Perlidae and Prosopistomatidae, which showed differences in occurrence between wet and dry periods, are known to occur in gravel, pebble and small stone-dominated habitats (Ogbogu 2006; Schletterer and Fureder 2009); so, any flow changes might have an impact on the stone biotope and hence on their occurrence and distribution.

Seasonal changes in macroinvertebrate assemblages have been found to have marked effects on many biotic indices (e.g. Reece et al. 2001; Sporka et al. 2006). Invertebrate life cycles and changes in macroinvertebrate composition may affect bioassessment metrics. Observed temporal variation in macroinvertebrate taxa and assemblages in this study have also influenced the ASPT in PHU and CEAL with combined biotopes, and the number of taxa, TARISS scores and ASPT in CEAL when stone biotopes were used. These marked differences between the wet and dry periods in TARISS reference metrics bring an alert to the use of reference conditions for comparison with test sites. Reference conditions should be able to differentiate natural temporal variation from variation caused by human disturbance and stressors. As a consequence, test sites from wet period should be compared to wet period reference conditions and test sites from dry period should be compared to dry period reference conditions. In this way, hydrology is taken into account, and potential biases associated with hydrological changes are accounted for.

Because of spatio-temporal variation in macroinvertebrate assemblages, Ormerod (1987) suggested that the most accurate data set of macroinvertebrate assemblages involves sampling that combines both habitat and seasonal data. Results from this study support the higher precision in using combined biotopes and sampling periods reference conditions for detecting changes in test sites. In all TARISS metrics, combined spatial and temporal reference data set resulted in higher significances of variation from test sites of all three sampling periods, suggesting that it is a more reliable and possibly accurate way of categorizing the types of macroinvertebrate data in river systems.

Understanding natural variability in macroinvertebrate taxa and TARISS metrics is important in monitoring programmes of river and stream ecosystems where a particular site is examined for changes over a period of time, i.e. annually. In such a situation, temporal changes can easily be misinterpreted for changes due to anthropogenic impacts. Thus, in regions where temporal variability is apparent, e.g. CEAL, monitoring programmes should be conducted over a period of time and seasonal reference conditions should be used to detect and interpret changes at a site.

It is recommended that, in order to account for temporal variability when using TARISS it is important to ascertain the presence of significant temporal variability in TARISS metrics among the main seasons in a particular region. If there are no significant temporal differences among seasons, then a combined-seasonal reference condition should be used. In cases where there are significant differences among seasons, it is best recommended to use season-specific reference conditions to detect changes in their respective test sites. However, even for sites where seasonal differences in TARISS metrics exist, combined reference metrics are also suitable for use in situations where the development of seasonal reference conditions is not feasible. The use of combined reference condition approach is useful in bioassessment, as one combined reference condition can be used to compare data from all sampling periods. Another advantage of a reference condition developed from multiple sampling periods is that it reduces challenges of deciding a suitable time period for sampling, since a test sample from any time of the year can be compared with a single set of reference conditions. Another option in regions where seasonal variations exist is to conduct bioassessment only during one season and use a single-season established reference condition.

Conclusion

Several countries in the Eastern and Southern Africa region have adopted the bioassessment methods in river health assessments. However, the successful application of these methods still requires the incorporation of the same into policies, acts and bylaws governing water resources management within respective authorities such as ministries and environment management councils and authorities. These policies should emphasize aquatic ecosystem protection and conservation as a key component in water resources and ecosystem management. Moreover, there is a need to establish and improve regional and international networking among researchers, stakeholders and governing authorities to improve biological monitoring through research and innovative practices that are ecologically oriented.

The networking should focus on developing one method and/or harmonizing existing methods that are applied in Africa; and should focus on a multi-indicator-based method. This should be followed by fostering two-way interactions between scientists on one hand, and the general public and decision makers, on the other, in order to fully realize the potential for bioassessment in the regions. In addition, there is a need for more studies on taxonomic clarification and ecological requirements of some bioindicators in the regions. This should help to resolve issues of abundant undescribed taxa, as in the case of the Manyame Catchment, Zimbabwe, which may lead to misinterpretation of water quality results. Reliable data sets on taxonomy and autecology of many bioindicators will improve the efficacy of the indicators as tools in river health monitoring in the region.

Finally, resources should be channeled towards training, and acquiring and maintaining the necessary infrastructure for biomonitoring in the region. Moreover, bioassessment methods should be blended into existing human resource's training programmes. In addition, bioassessment can be used as interface in primary and secondary educational activities, promoting the development of a constructive learning process in keeping with current educational trends based on concrete observations of the anthropogenic effects on aquatic environments.

Notes

1. A process of using biota or part of the biota or their responses to determine if biological properties of an ecosystem have been altered by human activities.
2. A systematic use of biota or part of the biota or their responses to determine condition or changes in the environment, usually part of quality control or management programmes.

References

Barbour, M.T., Gerritsen, J.B., Snyder, D.& Stribling J.B., 1999, Rapid Bioassessment Protocols for Use in Streams and Wadeable Rivers: Periphyton, Benthic Macroinvertebrates and Fish, Second Edition. EPA 841-B-99-002, Washington, D.C: U.S. Environmental Protection Agency, Office of Water.

Bartram, J. & Balance R., 1996, Water Quality Monitoring: A practical guide to the design and Implementation of Fresh Water Quality Studies and Monitoring Programmes, London: UNEP, Taylor & Frances.

Bere, T. & Tundisi, J.G., 2010a, 'Biological monitoring of lotic ecosystems: the role of diatoms', *Brazilian Journal of Biology*, 70, 493-502.

Bere, T. & Tundisi, J. G., 2010b, 'Epipsammic diatoms in streams influenced by urban pollution', *Brazilian Journal of Biology*, 70, 921-930.

Bere, T. & Tundisi, J.G., 2011, 'Influence of land-use patterns on benthic diatom communities and water quality in the tropical Monjolinho hydrological basin', São Carlos-SP, Brazil', Water, SA, 37, 93-102.

Bere, T. & Tundisi, J.G., 2011a, 'Influence of ionic strength and conductivity on benthic diatom communities in a tropical river (Monjolinho), São Carlos-SP, Brazil', *Hydrobiologia*, 661, 261-276.

Bere, T. & Tundisi, J.G., 2011b, 'Influence of land-use patterns on benthic diatom communities and water quality in the tropical Monjolinho hydrological basin, São Carlos-SP, Brazil', *Water*, SA, 37, 93-102.

Bere, T. & Mangadze, T., 2014, 'Diatom communities in streams draining urban areas: community structure in relation to environmental variables', *Journal of Tropical Ecology*, 55(2), 271-281.

Bere, T., Mangadze, T. & Mwedzi, T., 2014, 'The application and testing of diatom-based indices of water quality assessment in the Chinhoyi Town, Zimbabwe', *Water, SA*, 3, 503-511.

Bere, T. & Nyamupingidza, B., 2014, 'Use of biological monitoring tools beyond their country of origin: a case study of the South African Scoring System Version 5 (SASS5)', *Hydrobiologia*, 722, 223-232.

Biggs, B.J.F., 1989, 'Biomonitoring of organic pollution using periphyton, South Branch, Canterbury, New Zealand', *New Zealand Journal of Marine and Freshwater Research*, 23, 263-274.

Buffagni, A., Erba, S., Cazzola, M.& Kemp, L.L., 2004, 'The AQEM multimetric system for the southern Italian Alpennines: assessing the impact of water quality & habitat degradation on pool macroinvertebrates in Mediterranean rivers', *Hydrobiologia*, 516, 313-329.

CCME, 2006, 'Developing biocriteria as a water quality assessment tool in Canada: scoping assessment', Canada: Canadian Council of Ministers of the Environment, PN 1350, 53 pp.

Chakona, A., Phiri, C., Chinamaringa, A. & Muller, N., 2009, 'Changes in biota along a dry land river in north-western Zimbabwe: declines and improvements in river health related to land use', *Aquatic Ecology*, 4, 1095-1106.

Chakona, A., Phiri, C., Magadza, C.H.D. & Brendonck, L., 2008, 'The influence of habitat structure and flow permanence on macroinvertebrate assemblages in temporary rivers in northwestern Zimbabwe', *Hydrobiologia*, 607, 199-209.

Chapman, D., 1996, 'Water Quality Assessments-A Guide to Use of Biota, Sediments and Water in Environmental Monitoring-Second Edition, UNESCO/WHO/UNEP, Cambridge University Press communities', in Calow, P.& Petts G.E., eds, The Rivers Handbook, Volume 2: Hydrological and Ecological Principles. Oxford, UK: Blackwell Scientific Publications.

Chutter, F.M., 1995, 'The role of aquatic organisms in the management of river basins for sustainable utilisation', *Water Science and Technology* 32(5-6), 283-291.

Collares-Perreira, M.J. & Cowx, I.G., 2004, 'The role of catchment scale environmental management in freshwater fish conservation', *Fisheries Management and Ecology*, 2004,11, 303-312.

Daily, G.C., Söderqvist, T., Aniyar, S., Arrow, K., Dasguota, P., Ehrlich, P., Folke, C., Jansson, A., Jansson, B., Kautsky, N., Levin, S., Lunchenco, J., Mäler, K., Simpson, D., Starrett, D., Tilman, D. & Walker, B., 2000, 'The value of nature and the nature of value', *Science*, 289, 395-396.

Dallas, H.F., 2000, 'Ecological reference conditions for riverine macro invertebrates and the River Health Programme, South Africa', 1st WARFSA/ WaterNet Symposium: Sustainable Use of Water Resources, Maputo,1-2 November 2000.

Dallas, H.F. & Day J.A., 1993, 'The effects of water quality on riverine ecosystems: a review. Pretoria', South Africa: Water Research Commission Technical Report TT61/93, 240 pp.

Dallas, H. F., Kennedy, M., Taylar, J., Lowe, S., & Murphy., 2010, 'SAFRASS. South African Rivers Assessment Scheme', WP4, Review Paper.

Dallas, H.F., 2004, 'Spatial Variability in Macroinvertebrate Assemblages: Comparing Regional and Multivariate Approaches for Classifying Reference Sites in South Africa', *African Journal of Aquatic Science*, 29 (2), 161-171.

Davies, B.R. & Wishart, M.J., 2000, 'River Conservation in The Countries of The Southern African Development Community (SADC)', in P.J. Boon, B.R. Davies, & G.E. Petts, eds, Global Perspectives on River Conservation. Science, Policy And Practice, Chichester: John Whiley & Sons, pp. 179-204.

De La Rey, P.A., Taylor, J.C., Laas A., Van Rensburg, L. & Vosloo A., 2004, 'Determining The Possible Application Value Of Diatoms As Indicators of General Water Quality: A Comparison With SASS 5', *Water SA*, 30 (3), 325-332.

Descy, J.P. & Coste, M., 1991, 'A Test of Methods for Assessing Water Quality Based on Diatoms', *Verhandlungen Der Internationalen Vereinigung Für Theoretische Und Angewandte Limnologie*, 24(4), 2112-2116.

Dickens, C.W.S. & Graham, M., 2002, 'The South African Scoring System (SASS) Version 5. Rapid Bio-Assessment Method for Rivers', *African Journal of Aquatic Science*, 27, 1-10.

Ferreira, T., Caiola, N., Casals, F., Oliveira, J.M. & De Sostoa, A., 2007, 'Assessing Perturbation Of River Fish Communities in The Iberian Ecoregion', *Fisheries Management and Ecology*, 14, 519-530.

Giller, P.S. & Malmqvist, B., 1998, The Biology Of Streams And Rivers, New York: Oxford University Press.

Gratwicke, B., Marshall, B.E & Nhiwatiwa, T., 2003, 'The distribution and relative abundance of stream fishes in the upper Manyame River, Zimbabwe, in relation to land use, pollution and exotic predators', *African Journal of Aquatic Science*, 28, 25-34.

Gratwicke, B., 1998, 'An introduction to aquatic biological monitoring', *Zimbabwe Science News*, 32 (4), 75-78.

Gratwicke, B., 1999, 'The effect of season on a biotic water quality index: a case study of the Yellow Jacket and Mazowe Rivers, Zimbabwe', *Southern African Journal of Aquatic Sciences*, 24(1/2), 24-35.

Harding, W.R., Archibald, C.G.M. & Taylor, J.C., 2005, 'The relevance of diatoms for water quality assessment in South Africa: A position paper', *Water SA*, 31(1), 41-46. URL: http://www.wrc.org.za.

Harrison, A.D., 1966, 'Recolonisation of a Rhodesian stream after drought', Archiv fur *Hydrobiologie*, 62, 405-421.

Hein, L., Van Koppen, K., De Groot, R.S. & Van Ierland, E.C., 2006, 'Spatial scales, stakeholders and the valuation of ecosystem services', *Ecological Economics*, 57, 209-228.

Hellawell, J. M., 1986, Biological indices of freshwater pollution and management, London: Applied Science Publishers.

Hering, D., Johnson, R.K., Kramm, S., Schmutz, S., Szoszkiewicz, K. & Verdonschot, P.F.M., 2006, 'Assessment of European streams with diatoms, macrophytes, macroinvertebrates and fish: a comparative metric-based analysis of organism response to stress', *Freshwater Biology*, 51, 1757-1785.

Infante, D.M., Allan J.D., Linke S. & Norris R.H., 2009, 'Relationship of fish and mac-roinvertebrate assemblages to environmental factors: implications for community concordance', *Hydrobiologia*, 623, 87-103.

Johnson, R.K., Hering, D., Furse, M.T. & Verdonschot, P.F.M., 2006, 'Indicators of ecological change: comparison of the early response of four organism groups to stress gradients', *Hydrobiologia*, 566, 139-152.

Kadye, W.T., 2008, 'The application of a Fish Assemblage Integrity Index (FAII) in a Southern African river system', *Water SA*, 34, 25-32.

Kaaya, L.T., Jenny, D. & Dallas, H.F., 2015, 'Tanzania River Scoring System (TARISS): a macroinvertebrate-based biotic index for rapid bioassessment of rivers', *African Journal of Aquatic Sciences*, 40(2), 109-117. DOI: 10.2989/16085914.2015.1051941

Kaaya, L.T., 2015, 'Towards a classification of Tanzanian rivers: a bioassessment and eco-logical management tool. A case study of the Pangani, Rufiji and Wami-Ruvu river basins', *African journal of Aquatic Sciences*, 40:1, 37-45, DOI:10.2989/16085914.2015.1008970.

Karr, J.R., 1981, 'Assessment of biotic integrity using fish communities', Fisheries, 66, 21-71. Karr, J. R. & Chu, E. W., 2000, 'Sustaining living rivers', *Hydrobiologia*, 422/423, 1-14.

Karr, J. R., 1991. Biological integrity: a long-neglected aspect of water resource manage-ment. *Ecological Application*, 1, 66-84.

Karr, J. R., Fausch, K. D., Angermeier, P. L., Yant, P. R. & Schlosser, I. J. 1986. Assessing Biological Integrity in Running Waters-a method and its Rationale. Illinois Natural History Survey Special Publication, 5, 28 pp.

Karr, J.R. & Chu, E.W., 1999, Restoring Life in Running Waters: Better Biological Mon-itoring, Washington, DC: Island Press.

Karr, J.R. & Yoder, C.O., 2004, Biological Assessment and Criteria Improve Total maximum daily load decision making, *Journal of Environmental Engineering*, 130, 594-604.

Kleynhans, C.J., 1999, 'The development of a fish index to assess the biological integrity of South African rivers', *Water SA*, 25, 265-278.

Lobo, E.A., Callegaro, V.L.M., Hermany, G., Gomez, N. & Ector, L., 2004, 'Review of the use of microalgae in South America for monitoring rivers, with special reference to diatoms', *Vie et milieu*, 54(2-3), 105-114.

Lorenz, A., Hering, D., Feld, C.K. & Rolauffs, P., 2004, 'A new method for assessing the impact of hydromorphological degradation on the macroinvertebrate fauna of five German stream types', *Hydrobiologia*, 516, 107-127.

Magadza, C. H. D. & Masendu, H., 1986, 'Some Observations On Mine Effluent In the yellow jacket stream, Zimbabwe', *The Science News*, 20, 11-28.

Mangadze, T., Bere, T. & Mwedzi, T., 2016, 'Choice of biota in stream assessment and monitoring programs in tropical streams: a comparison of diatoms, macroinverte-brates and fish', *Ecological Indicators*, 63,128-143.

Menezes, S., Baird, D. J. & Soares, A.M.V.M., 2010, 'Beyond taxonomy: a review of macroinvertebrate trait-based community descriptors as tools for freshwater biomoni-toring', *Journal of Applied Ecology*, 47, 711-719.

Metcalfe, J.L., 1989, 'Biological Water Quality Assessment of Running Waters Based on Macroinvertebrate Communities: History and Present Status in Europe', *Environmental Pollution*, 60, 101-139.

Metcalfe-Smith, J.L., 1994, 'Biological water quality assessment of rivers: use of macroinvertebrate communities', in P. Calow, & G. E. Petts, eds, The Rivers Handbook, Vol. II. London: Blackwell Scientific Publications, 144-170.

Minshull, J.L., 1993, 'How do we conserve the fishes of Zimbabwe?' *Zimbabwe Science News*, 27, 90-94.

Moyo, N.A.G. & Phiri, C., 2002, 'The degradation of an urban stream in Harare', *African Journal Ecology*, 40, 401-406.

Moyo, N.A.G. & Worster, K., 1997, 'The effects of organic pollution on Mukuvisi River, Harare', in N.A.G. Moyo, ed., Lake Chivero: a polluted lake, Harare, Zimbabwe: University of Zimbabwe Publications.

Mtetwa, S., 2001, 'Assessment of reference sites for a biomonitoring water quality assessment network in Zimbabwe', Volume 1: Initiation of a Biomonitoring Programme in Zimbabwe.

Ndebele-Murisa, M. R., 2012, Biological monitoring and pollution assessment of the Mukuvisi River, Harare, Zimbabwe. *Lakes and Reservoirs; Research and Management*, 17, 73-80.

Niemi, G.J. & Mcdonald, M.E., 2004, 'Application of ecological indicators', *Annual Review of Ecology, Evolution, and Systematics*, 35, 89-111.

Ormerod, S.J., 2003, 'Restoration in applied ecology: editor's introduction', *Journal of Applied Ecology*, 40, 44-50.

Pan, Y., Stevenson, R.J., Hill, B.H., Herlihy, A.T.& Collins, G.B., 1996, 'Using diatoms as indicators of ecological conditions in lotic systems: a regional assessment', *Journal of North American Bethological Society*, 15, 481-495.

PBWO/IUCN, 2007, 'River Health Assessment Report. Moshi, Pangani River Basin Flow Assessment', 121pp.

Phiri, C., 2000, 'An assessment of the health of two rivers within Harare, Zimbabwe, on the basis of macroinvertebrate community structure and selected physicochemical variables', *African Journal of Aquatic Science*, 25, 134-141.

Phiri, C., 1998, 'A Comparative study of the Water Quality of the Gwebi and Mukuvisi Rivers', Unpublished MScThesis, Department of Biological Sciences, Harare, Zimbabwe: University of Zimbabwe.

Phiri, C., Day, J., Chimbari, M & Dhlomo, E., 2007, 'Epiphytic diatoms associated with a submerged macrophyte, Vallisneria aethiopica, in the shallow marginal areas of Sanyati Basin (Lake Kariba): a preliminary assessment of their use as biomonitoring tools'. *Aquatic Ecology*, 41, 169-181.

Pont, D., Hugueny, B., Beier, U., Goffaux, D., Melcher, A., Noble, R., Rogers, C., Roset, N. & Schmutz, S., 2006, 'Assessing river biotic condition at a continental scale: a European approach using functional metrics and fish assemblages', *Journal of Applied Ecology*, 43, 70-80.

Ravengai, S., Love, D., Love, I., Gratwicke, B., Mandingaisa, O. & Owen, R.J.S., 2005, Impact of Iron Duke Pyrite Mine on water chemistry and aquatic life-Mazowe Valley, Zimbabwe, Water SA, 31(2), 119-228.

Reynoldson, T.B. & Metcalfe-Smith, J.L., 1992, 'An overview of the assessment of aquatic ecosystem health using benthic macroinvertebrates', *Journal of Aquatic Ecosystem Health*, 1, 295-308.

Rosenberg, D. M. & Resh, V. H., 1993, Freshwater Bio-monitoring and Benthic Macro-invertebrates, New York, USA: Chapman and Hall.

Roux, D.J., 1997, 'National Aquatic Ecosystems Biological monitoring Programme: Overview of the design process and guidelines for implementation', NAEBP Report Series 6. Pretoria. South Africa: Institute of Water Affairs and Forestry, 113 pp.

Sandin, L., Dahl, J. & Johnson, R.K., 2004, 'Assessing acid stress in Swedish boreal and alpine streams using benthic macroinvertebrates', in D. Hering, P.F.M. Verdonschot, O. Moog, & L. Sandin, eds, 'Integrated Assessment of Running Waters in Europe'. *Hydrobiology*, 516, 129-148.

Statzner, B., Bis, B., Dolédec, S. & Usseglio-Polatera, P., 2001, 'Perspectives for biomonitoring at large spatial scales: a unified measure for the functional com-position of invertebrate communities in European running waters', *Basic Applied Ecology*, 2, 73-85.

Taylor, J.C., Prygiel, J., Vosloo, A., Pieter, A., Rey, D. & Ransburg, L.V., 2007, 'Can diatom based pollution indices be used for biomonitoring in South Africa? A case study of the Crocodile West and Marico water management area', *Hydrobiologia*, 592, 455-464.

Waite, I. R., Herlihy, A. T., Larsen, D. P & Klemm, D.J., 2000, 'Comparing strengths of geographic and non-geographic classifications of stream benthic macroinvertebrates in the Mid-Atlantic Highlands, USA', *Journal of the North American Benthological Society*, 19, 429-441.

Walsh, C.J., 2006, 'Biological indicators of stream health using macroinvertebrate assemblage composition: A comparison of sensitivity to an urban gradient', *Marine and Freshwater Research*, 57, 37-47.

Whitton, B.A. & Rott, E., 1996, 'Use of Algae for Monitoring Rivers', Ii. Proceedings of 2nd European Workshop, Innsbruck, 1995, Innsbruck:Universität Innsbruck, 196 pp.

Yoder, C. O. & Rankin, E. T., 1998, 'The role of biological indicators in a state water quality management process', *Environmental Monitoring and Assessment*, 51, 61-88.

Zalewski, M., 2002, 'Ecohydrology-the use of ecological and hydrological processes for sustainable management of water resources', *Hydrological Sciences Journal*, 47, 825-834.

5

Historical and Future Climate Scenarios of the Zambezi River Basin

Masumbuko Semba, Mzime Ndebele-Murisa,
Chipo Plaxedes Mubaya, Ismael Aaron Kimirei,
Geoffrey Chavula, Tongayi Mwedzi,
Tendayi Mutimukuru-Maravanyika and Sandra Zenda

Introduction

Africa faces a plethora of challenges and chief among these is a change in the climate (Zakaria and Maharjan 2014) which is one of the key factors affecting the ecology and hydrology of its river basins (Kusangaya et al. 2014). Beilfuss (2012) proposed that Africa's arid regions are highly vulnerable to climate change with the Zambezi River Basin (ZRB) being particularly at risk (Kling et al. 2014). After the Nile and Niger rivers, the ZRB is the next most trans-boundary river basin in Africa as it serves eight African countries. Consequently, water resource development planning is crucial, since any changes in climate will impact the hydrological cycle and the amount of water retained in hydrological systems (Beilfuss 2012) of which only up to 3 per cent is readily available as usable and shared freshwater. Like some Sub-Saharan countries, which have experienced up to 0.5°C increases in temperature (Hendrix and Glaser 2007), the Zambezi River Basin is also facing changes in climate (Ndhlovu 2013). A recent study by Kling et al. (2014) reported rises in temperature and more variable precipitation in the basin since the 1980s. Such historical climatic changes, and those projected towards the mid-century (2050), are of concern with serious social and economic implications to local communities (Mubaya et al. 2012).

The Intergovernmental Panel on Climate Change (IPCC) projected a global decadal temperature rise of 0.2°C (IPCC 2007). However, regional climate projections for southern Africa have reflected a much higher warming of 2°C by the mid-century (2050) and 3°C increase by the end of the century (Schaeffer et al. 2015). This projection aligns well with the findings of Ndebele-Murisa et al. (2013), which reflected that annual temperatures in the Middle Zambezi Basin (Gwembe Valley) have increased by 3.4°C between 1963 and 2000. It is evident that the ZRB experiences much higher warming than the predicted 0.78°C for Zimbabwe, and 1.6 to 2.5°C (IPCC 2001) for the semi-arid regions (Hulme et al. 2001; IPCC 2007). A comparison of estimated changes in precipitation, potential evapotranspiration and runoff in several major river basins in Africa (Table 5.1) suggests that the ZRB is severely impacted by climate change (Arnell 1999; Desanker and Justice 2001; Desanker and Magadza 2001; Hulme et al. 2001; Easterling and Apps 2005; Hulme 2005; Shongwe et al. 2009; Tauya 2010; Ndebele-Murisa et al. 2011; Schefuss et al. 2011; Beilfuss 2012; Kusangaya et al. 2014).

A study by Beilfuss (2012) indicated imbalance between mean annual potential evapotranspiration and mean annual precipitation in the ZRB. The mean annual potential evapotranspiration of 1,560 mm experienced in ZRB significantly exceeds the mean annual precipitation. As a consequence, elevated temperatures are expected to increase annual evapotranspiration rates profoundly (Arnell 2004). These rates are estimated to increase by 10 per cent to 25 per cent between 2000 and 2100 (Arnell 2004) with much of the basin's surface becoming drier (Tauya 2010; Pricope and Binford 2012; Kling et al. 2014). This will reduce the amount of water available in the basin for social, economic and ecological functions, thereby severely impacting rural communities whose livelihoods depend on the basin (Dugan 1993; Kling et al. 2014).

Table 5.1: Estimated percentage changes in runoff, precipitation and potential evaporation for ten river basins in Africa

Basin	Percentage Change		
	Precipitation	Potential Evaporation	Run-off
Nile	10	10	0
Niger	10	10	10
Volta	0	4-5	0-15
Zaire	10	10-18	10-15
Ogoonue	-02 to 20	10	-20 to 25
Rufiji	-10 to 20	20	-10 to 25
Zambezi	-10 to 20	10 to 25	-26 to-40
Ruvuma	-10 to 5	25	-30 to-40
Limpopo	-5 to-15	5 to 20	-25 to-35
Orange	5 to 5	4 to 10	-10 to 10

Source: IPCC (2001)

Desanker and Magadza (2001) projected that a 20 per cent decrease in precipitation and a 25 per cent increase in evaporation, combined with increased development and water extraction across the Zambezi River Basin, would result in a 26 to 40 per cent decline in water runoff with consequent reductions in stream flow. Indeed, projected reductions in stream flow of 5 to 10 per cent (Kling et al. 2014) have been associated with increased evaporation and transpiration rates resulting from a 1°C rise in temperature across the Zambezi River Basin (Chenje 2000). These projections match with projected reduction of 11, 23, and 30 per cent in stream flow for 2°C, 3°C and 4°C increases in temperature in other hydrological systems (Henson 2011). The current shortage of hydroelectric power across the southern African region has also been contributed by reduced precipitation, which has led to decreasing stream flows and water levels in some rivers, resulting in increased frequency of power rationing, particularly within the Zambezi Basin where some of the major hydroelectric plants (Kariba, Cabora Bassa, Kafue and Shire) are located.

In view of the above-mentioned anticipated changes in climate, this chapter analyses historical and projected future trends in temperature and precipitation across the Zambezi River Basin. The chapter also reviews the roles of Local Indigenous Knowledge Systems and Practices (LIKSP) in order to integrate these with interventions and adaptation planning for climate change. To address how climate varies across the Zambezi River Basin, the study uses five major river catchments as case studies and assesses the potential impacts of climate change on the natural and agricultural ecosystems on whose resources human livelihoods depend. The five river catchments include the Barotse Flood plain in Zambia, Manyame in Zimbabwe, the Shire in Malawi, Ruhuhu and Songwe in Tanzania, and represent 38 per cent of the Zambezi River Basin's 13 catchments. The catchments were selected based on their spatial distribution in the basin and the availability of historical and modelled climate data.

Climate Change Impacts on Livelihoods and Ecosystems in Africa

In Africa, agriculture is the most important economic sector. It accounts for 65 per cent of the continent's employment (Haggblade and Hazell 2010). While domestic trade accounts for 75 per cent and about 15 per cent of its total annual Gross Domestic Product (GDP) worth US$100 billion, agriculture has the largest share of more than 40 per cent of Africa's total export earnings (Jayaram et al. 2010; Ringler 2010; World Bank 2012; FAO 2014; World Bank 2015). Mubaya (2006) noted that communal farmers, who constitute the largest percentage of farmers in most African countries, use both agricultural and non-agricultural strategies as their livelihood systems, but agriculture contributes most to their livelihood. Cultivated land in Sub-Saharan Africa (SSA) for instance, has increased from 166 million ha in 1970 to 202 million ha in 1999 of which 192 million ha is rain-fed (Saghir 2014). This rain-fed cultivation is largely vulnerable to climate change (Desanker and Justice 2001), a consequence of which in tropical agro-ecological systems is a significant reduction in crop and rangeland production and an isolation of marginal lands from agricultural cultivation.

Gitay et al. (2001) calculated that Africa could lose about 12 per cent in its net crop production as a result of climate change. However, a more recent study reported an even higher potential loss of net crop production of up to 40 per cent to some Sub-Saharan Africa countries (Mwingira et al. 2011), which translates into a loss of US$10-60 billion, valued at 1990 prices. The distribution of these losses is not uniform in the region as certain countries are affected more than others. For instance, it is projected that by 2100 Zambia, Niger and Chad will lose practically all of their entire farming sector while 70 per cent of Burkina Faso, Togo, Botswana, Guinea Bissau, and Gambia's farming output could be wiped out by climate impacts (Mendelsohn et al. 2000). In fact, food production is on the decline in most of Sub-Saharan Africa and is not keeping in pace with population increase (Schmidhuber and Tubiello 2007; Bourgeois-Pichat 2008).

Like agricultural systems, the productivity, structure and functioning of natural ecosystems is also driven primarily by the climate (Gaitan et al. 2014). Precipitation, for instance, drives plant growth in the terrestrial environment, which in turn determines, to a large extent, plant biomass (Li et al. 2014), and as a consequence, primary production. Similarly, primary production in aquatic ecosystems is also largely driven by climate factors such as temperature, wind dynamics and precipitation (O'Reilly et al. 2003; Wrona et al. 2006; Ndebele-Murisa et al. 2012), whereas in terrestrial systems it is largely dependent on available soil moisture and nutritional status which in turn is influenced by the nature of parent rocks, soil physical properties (Abbott and Murphy 2007) and human inputs of nutrients in the form of nitrogenous and phosphorous compounds (Bureau and Hua 2010; see Chapters 3 and 4). Both available moisture and nutrients influence aquatic primary producers such as periphyton, algae and diatoms aquatic macrophytes, riparian vegetation and mangrove forests as well as other trees, shrubs, herbs, grasses, sedges and reeds which in turn, determine the biomass of successive trophic levels that include secondary and tertiary consumers in ecological systems (Shurin et al. 2012; Agrawal and Gopal 2013). In terrestrial ecosystems the successive trophic levels include herbivores such as ungulates, as well as insects and some fungi, while in aquatic ecosystems, such as in the Zambezi waters, they include zooplankton, insects, fish and humans which feed on and utilize the diverse aquatic primary producers and their bi-products.

Apart from precipitation, temperature directly affects metabolic processes of all living organisms (Ohlberger et al. 2012) and is an important factor determining ecosystem health and functioning. All living organisms function within a certain range of temperature beyond which productivity, breeding and reproduction can be impaired or cease completely (Wallenstein and Hall 2012). These temperature thresholds are crucial for species survival and maintenance of biological diversity on which some major economies and tourism depend on in Sub-Saharan Africa. For instance, the Okavango River Basin and its Delta, a RAMSAR site, possess a high diversity of birds and insects, which attract large numbers of tourists (Mopelwa and Blignaut 2014). Similarly, the species of cichlid (Cichlidae family) fish endemic to Lake Malawi (Weyl et al. 2010) are a major tourist attraction. Shifts of biomes due to climatic change have been projected for much of the southern African region with as much as a 4°C rise in temperature and a 15 per cent decline in precipitation with the exception of the East African region where wetter conditions and increased incidence of floods are projected (Conway 2009). Such changes are likely to affect many of the delicate ecosystems and biological diversity of the region and consequently the livelihoods that depend on ecosystem goods and services.

Climate in the Zambezi River Basin

The Zambezi River Basin climate varies with space and time because of the interaction of regional and global climate systems and diverse terrains across the eight countries proximate to the basin. The Basin's elevations range from 0 meters above sea level at the delta, where the river meets the Indian Ocean in Mozambique to 2,900 meters above sea level in Songwe, Tanzania (Figure 5.1a; also see Chapter 2's introduction). This wide range in elevation across the Zambezi River Basin (ZRB) results in wide spatial variations in temperature (Figure 5.1b) and precipitation (Figure 5.1c). The mean annual temperature ranges from 19°C at the highest elevation to 31°C at the lowest elevation (Figure 5.1b). Temporal variations in precipitation across the basin lead to remarkable differences of runoff from year to year (SADC 2007). On average, the annual precipitation across the ZRB ranges from 500 mm in the extreme southern and south-western parts of the basin to more than 1,400 mm in the Upper Zambezi and Kabompo sub-basins, around the north-eastern shores of Lake Malawi/Nyasa/Niassa in Tanzania, and in the southern border area between Malawi and Mozambique (Figure 5.1c).

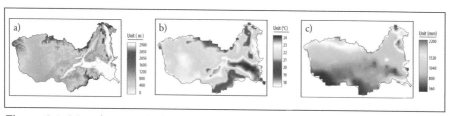

Figure 5.1: Map showing a) elevation, b) historical 30-year return mean temperature, and c) mean precipitation across the Zambezi River Basin

Analysis of Five Sub Catchments

We analysed historical precipitation and temperature data records for five selected Zambezi River catchments supplied. Historical data from the National Meteorological Agencies of Malawi, Tanzania, Zambia, and Zimbabwe (Table 5.2). As the temperature and precipitation records were non-normally distributed, skewed and incomplete, a non-parametric Mann-Kendall test was applied to test for seasonal and annual trends (Scott and Gemmell 2012; Millard 2013) and a Kruskal-Wallis test was used to test for significant decadal changes in temperature and precipitation for five selected catchments in the Zambezi River Basin from 1980 to 2014. The tests deal effectively with incomplete records (Millard 2013) and generating probability estimates that assessed significant increase or decrease trends in temperature and precipitation in the five selected river catchments in the Zambezi River Basin.

The annual and seasonal trends in maximum daily temperature for the selected Zambezi River catchments are presented in Table 5.3. All five catchments had significant seasonal trends in maximum temperature ($p < 0.05$). The Songwe catchment had the highest seasonal trends in maximum daily temperature with a magnitude of 0.3°C followed by the Shire, Manyame, Ruhuhu and Barotse catchments (Table 5.3). However, the Manyame, Shire and Songwe catchments showed a significant annual trend of maximum daily temperature (Figure 5.2). The Songwe reflected the highest annual increasing trend, (tau = 0.127), followed by the Shire then the Manyame (Table 5.3).

Table 5.2: Seasonal mean precipitation, minimum and maximum daily temperatures for five river catchments in the Zambezi River Basin between 1980 and 2014

Catchment	Country	Seasons	Months	Rain (mm)	Tmin (°C)	Tmax (°C)	Reference source
Barotse	Zambia	hot wet	Nov-Apr	716	15.9	30.8	ZMD (1987-2013)
		cool dry	May-Jun	-	-	-	-
		hot dry	Aug-Oct	-	-	-	-
Manyame	Zimbabwe	Summer	Nov-Apr	768	13.2	26.2	ZMS (1980-2014)
		Winter	May-Oct	-	-	-	-
Shire	Malawi	Wet	Nov-Mar	1094	17.8	30.4	MMS (1980-2009)
		Dry	Apr-Oct	-	-	-	-
Songwe	Tanzania	Wet	Nov-May	903	11.0	24.2	TMA (1980-2014)
		Dry	Jun-Oct	-	-	-	-
Ruhuhu	Tanzania	Wet	Nov-May	1047	16.0	27.2	TMA (1980-2014)
		Dry	Jun-Oct	-	-	-	-

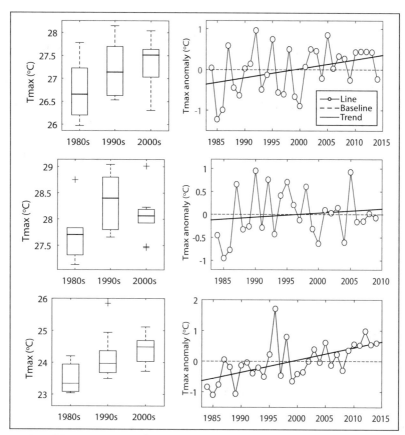

Figure 5.2: Maximum annual temperature anomalies for Manyame (top), Shire (middle) and Songwe (bottom) showing significant increasing trends (line plots) and decadal temperature change (boxplots)

Table 5.3: The seasonal and annual trends in maximum air temperatures in five catchments within the Zambezi River Basin

River Catchment	Seasonal Trend			Annual Trend		
	tau	*z*	*p*	*tau*	*z*	*p*
Barotse	0.07	3.07	0.002	0.005	0.27	0.787
Manyame	0.15	7.30	0.000	0.058	2.86	0.004
Ruhuhu	0.11	3.15	0.002	0.026	0.79	0.432
Shire	0.20	6.74	0.000	0.072	2.63	0.009
Songwe	0.30	8.54	0.000	0.127	3.74	0.000

Similar to maximum temperature, the seasonal trend analysis showed that minimum temperatures have increased significantly over the past thirty-four years for all the five selected catchments in the ZRB (Table 5.4). The Mann-Kendall trend analysis indicates a significant seasonal increasing trend in minimum temperatures which ranged between 0.07 and 0.34°C ($p < 0.05$) for all five basins, with Shire and Songwe catchments, located on the north eastern side of the basin, having higher trends of minimum temperature (Table 5.4). The minimum temperatures in the Barotse, Shire and Songwe catchments showed significant increasing annual trends ($p \leq 0.01$), the trends for Manyame and Ruhuhu catchments were insignificant ($p \geq 0.05$; Table 5.4).

Table 5.4: The seasonal and annual trends in minimum air temperature for five river catchments within the Zambezi River Basin

River Catchment	Seasonal Trend			Annual Trend		
	tau	*z*	*p*	*tau*	*z*	*p*
Barotse	0.16	6.33	0.00	0.07	2.70	0.01
Manyame	0.07	3.57	0.00	0.03	1.54	0.12
Ruhuhu	0.17	4.95	0.00	0.05	1.49	0.14
Shire	0.34	12.04	0.00	0.11	3.92	0.00
Songwe	0.26	7.38	0.00	0.09	2.70	0.00

Figure 5.3 shows catchments with increasing trends in minimum temperature in the Zambezi River Basin over the study period. The Kruskal-Wallis test indicated a significant decadal increase in minimum temperature for all the five catchments in the Zambezi River Basin over the last 34 years (X^2 (3) = 92.98, p <0.05).

Table 5.5 shows seasonal and annual increasing and decreasing trends in precipitation and their statistical significance for five catchments in the ZRB. The increasing seasonal trend in precipitation for Ruhuhu (*tau* = 0.01) and Barotse (*tau* = 0.03) mm catchment were insignificant ($p \geq 0.01$), whereas Manyame (*tau* =-0.02), Shire (*tau* =-0.03) mm and Songwe (tau =-0.03) had insignificant decreasing trends in precipitation ($p \geq 0.01$; Table 5.5). Coincidentally, the Manyame, Shire and Songwe catchments also had significant increases in annual maximum temperature but with Barotse replacing Manyame for annual minimum temperature (see Figure 5.2 and Figure 5.3). The Kruskal-Wallis test indicated that the observed decadal difference in the amount of precipitation observed in all catchments in the basin was insignificant (x^2 (2) = 0.83, p>0.05) (Figure 5.4). In general, the results indicate decreases in precipitation-albeit statistical insignificance over the Zambezi River Basin for the last three decades. The annual decreasing rate in precipitation in the basin ranged from 0.6 mm in

Manyame to 2.0 mm in Songwe and 1.4 mm in Shire. Nevertheless, the seasonal decreasing trend was higher than the annual trend in all five catchments (Table 5.5). Conversely, the annual and increased trend in annual precipitation varied from 0.1 in Ruhuhu and 2.2 mm in Barotse.

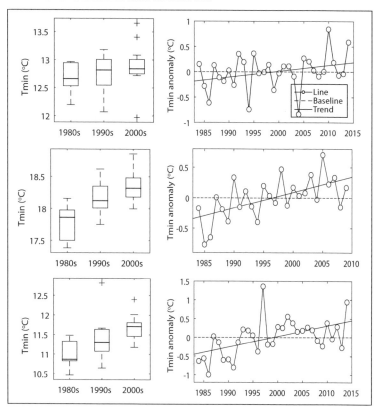

Figure 5.3: Minimum temperature anomalies for Manyame (top), Shire (middle) and Songwe (bottom) showing significant increasing trends (line plots) and decadal temperature change (boxplots)

Table 5.5 Seasonal and annual trends in precipitation for five catchments within the Zambezi River Basin

River Catchment	Seasonal Trend			Annual Trend		
	tau	*z*	*p*	*tau*	*z*	*p*
Barotse	0.03	1.48	0.14	0.022	0.99	0.32
Manyame	-0.02	-1.11	0.27	-0.006	-0.30	0.76
Ruhuhu	0.01	0.20	0.84	0.001	0.04	0.97
Shire	-0.03	-1.00	0.32	-0.014	-0.58	0.56
Songwe	-0.03	-0.97	0.33	-0.020	-0.61	0.54

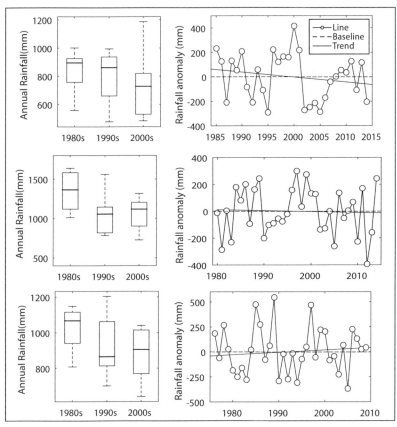

Figure 5.4: Precipitation anomalies (line plot) for three river catchments (Top = Manyame; Middle = Shire; Bottom = Songwe) in the Zambezi River Basin and the corresponding decadal precipitation variations (boxplots)

Scenarios and Projections of Climate Change in the Zambezi River Basin

The Intergovernmental Panel on Climate Change (IPCC) is a scientific intergovernmental body under the auspices of the United Nations. It is a legitimate international body mandated to analyse climate dynamics, whether natural or human-induced, and their impacts. In addition, though not prescriptive, the IPCC recommends options for adaptation and mitigation to climate change and variability (IPCC 2014b). The IPCC provides Global Circulation Models (GCMs) that are used to project future climate deviations. GCMs are invaluable tools for assessing potential climate changes resulting from increased greenhouse gas concentrations (Caesar et al. 2015). While global models are not specifically designed to simulate regional climate like for the Zambezi River Basin, the global physical consistency in GCMs along with the large number of simulations makes them useful (Tauya 2010). Despite their usefulness as some indication of possible future scenarios of climate, it is prudent to note that there is a level of uncertainty that comes these with climate model projections which is rarely quantified.

Uncertainty arises from standard errors in the model structure, scenarios and initial conditions and the response of the terrestrial carbon cycle, which renders some unreliability to the projections of impacts (Curry and Webster 2011; Ahlström et al. 2013). Hawkins and Sutton (2009) describe three main contributors of uncertainty in climate predictions, viz. (1) the concentration of greenhouse gases and other contributors to climate change, (2) the uncertainty in the response of the climate system, and (3) the model initial condition uncertainty. We therefore use a cautionary approach in employing GCMs for the assessment of future climate scenarios of the ZRB, cognizant of the underlying uncertainties.

Six emission scenarios are used to simulate changes in climate for mid-century and far future years (2046-65 and 2081-2100) relative to the historical period of 1981-2000 (IPCC 2014a). The global scenarios of change for both temperature and precipitation from the climate models for 2050, which is considered a median for the near future (2045-2065) term vary greatly and there is a lot of uncertainty (Kusangaya et al. 2014). For this study, we used the A1B and B1 scenarios. The A1B works with slightly warmer, and slightly wetter conditions with emphasis on balanced energy sources (IPCC 2014a). These would have more modest effects on ecosystems and agriculture over the next few decades. The B1 scenario projects hotter, drier outcomes that would lead to much larger negative impacts. These scenarios represent the range of outcomes projected by the climate models for average temperature and precipitation in the whole Zambezi River Basin area.

It is worth noting that there are, of course, a large number of intermediate outcomes that fall between these two scenarios. In addition, the projected changes in precipitation and temperature over the Zambezi River Basin were analysed based on the A1B climate scenario (IPCC 2007, 2014b). This is because it is expected that, in the future, emissions will increase as opposed to the decrease (B1) and the business as usual (A2) scenarios over SSA. In addition, we also used scenarios data generated from the Geophysical Fluid Dynamics Laboratory Coupled Model (GFDL-CM2.0). We chose this model because it was the only model that could generate temperature and precipitation data for both historical and future projections at medium emission. In our analysis, projected mid-century climate is compared to the corresponding historical mean climatology-the predicted changes in years 2046-2065 are calculated relative to the 1961-1990 baseline period following the standard for the World Meteorological Organization. It should be noted that the model simulations described here are from free-running climate model experiments. No observational data from the atmosphere, ocean, or land were ingested to force the model towards observations. Carbon dioxide and other anthropogenic greenhouse gas levels are prescribed in these models—they are not prognostics variables. The model generated its own internal variability and projected changes in temperature and precipitation for the Zambezi Basin.

Temperature

The mean surface temperature of historical (1961-1990), projected mid-century (2046-2065), and mean changes of temperature between the two time periods are shown in Figure 5.5. Positive values in Figure 5.5(c) indicate spatial increase of temperature in the basin, with ranges from 2.08 to 2.97°C. It is evident that the temperature changes reflect a spatial pattern across the basin (Figure 5.5c). Maximum values are distributed over the western and southern regions of the basin to the north and east and minimum values over much of central and north-western side of the basin. While regional projections of temperatures in arid regions matches the global ones, precipitation projections show more variations among models in the amplitude and direction of change, but in all models precipitation changes are enhanced as global temperature increases, and the degree of response is regionally specific (Kusangaya et al. 2014). This trend was observed for the historical analyses presented earlier, particularly for the Shire and Songwe catchments and, to a lesser extent, the Barotse and Manyame in the present study.

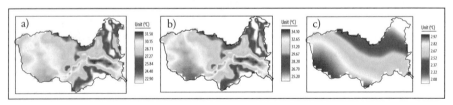

Figure 5.5: Multi-model mean temperature in (a) historical, 30-year return and (b) mid-century and (c) change between historical and simulated temperature by the GFDL-CM2.0 model in the SRES A1B scenario for the mid-century period (2046-2065)

Precipitation

The spatial distribution of precipitation for historical and mid-century is shown in Figure 5.6. The precipitation projection from the GFDL-CM2.0 model matches the spatial distribution across the basin with historical data. The northern region of the basin receives the highest amount of precipitation (Figure 5.6a). However, the basin will likely receive lower amounts of precipitation in the future (Figure 5.6b). Notwithstanding, the central and northern parts of the basin will likely experience modest increases in precipitation during the mid-century period (2046-2065), while the south-eastern and north-western regions will be drier than average (Figure 5.6c). The projected climate changes described here are driven primarily by increases in atmospheric greenhouse gases, while there may be some low frequency long-term natural cycle in influence as well. Mean precipitation will likely increase in the central and north-eastern region of the basin, while it is projected to decrease in the north-western and south-eastern region. It is projected that the proportion of drier areas will surpass the wet area in the mid-century. These findings correspond well with the projected changes in temperature in the basin (see Figure 5.5c).

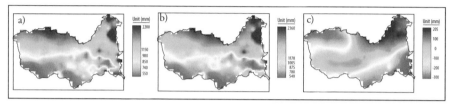

Figure 5.6: The multi-model mean precipitation in (a) historical 30-year return, (b) projected mid-century precipitation and (c) change between historical and mid-century precipitation as simulated by the GDLR2.0 model in the SRES A1B, mid-century (2046-2065) scenario

The projected climate in the Zambezi River Basin varies depending on different scenarios and models proposed by the IPCC (IPCC, 2014c). Projected changes in mean precipitation and temperature under A1B scenario for selected catchments in the Zambezi River Basin are summarized in Table 5.6. The A1B scenarios indicate a warmer future in the mid-century, with a climate sensitivity of above 2.4°C for minimum temperature and 2.1°C for maximum temperature in all the five catchments. The increase in minimum temperature is expected to range from 2.43°C in Ruhuhu to 2.92°C in Barotse and maximum temperature is projected to range in increase from 2.12°C in the Songwe to 2.79°C in the Barotse catchments (Table 5.6). The increase in the basin mean surface temperature by the mid-century (2045-2065) relative to 1961-1990 is likely to be 1.7°C to 2.03°C under A1B, and 2.43°C under the A2 scenario (Table 5.6).

In contrast to consistent patterns of increasing minimum and maximum temperature for the five catchments, trends in projected rainfall differ from one catchment to another, with increasing trends in some and decreasing trends in other sub-catchments (Figure 5.7). The decrease in rainfall is expected to range from-153 mm in the Shire catchment to about-6 mm in Manyame (Table 5.6). While the increase in rainfall is expected in the Barotse (70 mm), Ruhuhu (134 mm) and Songwe catchments (170 mm), it is expected to be highly variable in the Songwe River catchment (Table 5.6). In general, the SRES A1B projects increase in annual precipitation in mid-century for all catchments with the exception of the Shire and Manyame catchments (Figure 5.7a). The Barotse is projected to experience the highest increase of minimum (2.92°C) and maximum (2.72°C) temperature but moderate increase of precipitation (69 mm). In contrast to Barotse, the Ruhuhu and Songwe catchment are projected to experience low increase in temperatures but the highest increase in precipitation (Table 5.6). The Shire catchment is projected to experience increase in minimum and maximum temperature of 2.46 and 2.53°C, respectively, and is the only catchment projected to experience significant reduction in the amount of rainfall (Table 5.6).

Table 5.6: Statistic metrics of change in temperature and precipitation in the five river catchments under AIB scenarios

River Catchment	Change in climate variables (Mean ±SD)		
	Tmin (°C)	Tmax (°C)	Rainfall (mm)
Barotse	2.92 ±0.04	2.79 ±0.04	69.76 ±21.72
Manyame	2.63 ±0.04	2.66 ±0.03	-5.61 ±38.24
Ruhuhu	2.43 ±0.01	2.12 ±0.02	134.82±72.53
Shire	2.46 ±0.04	2.53 ±0.05	-153.93±63.02
Songwe	2.49 ±0.02	2.12 ±0.02	170.53±159.47

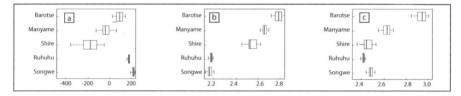

Figure 5.7: Boxplot of change in (a) rainfall, (b) maximum, and (c) minimum temperature for five sub-catchments in the Zambezi River Basin in mid-century (2046-2065) using the SRES A1B scenario. The boxplots indicate the central 50 per cent median and range of temperature and precipitation. The whiskers extend to the lower and upper values

Role of LIKSP as an Early-Warning System Under Climate Change Scenarios

More often than not, various disciplines within climate sciences work in silos and have not been able to integrate the different components which together make up today's climate. The physical sciences in particular, though they consider and input 'human contribution' to the climate problem in their studies, mostly concentrate on hardcore physical sciences (physical geography, physics, mathematics, statistics, and to a lesser extent geology, chemistry, ecology and biology) at the expense of social sciences (political sciences, geography, communication, anthropology, psychology, policy and law) and engineering (fuels and energy, civil, mechanical), and health sciences (epidemiology, medicine) among others. This presents a disjoint in the approach to solving the problem(s) of climate change and variability, which are extremely cross cutting in nature. In fact, the nature of climate itself is that it 'knows no boundary' and therefore a holistic approach, which is multi-disciplinary, is the more appropriate way of going about this quandary.

In this chapter, we attempt to bridge the gap described earlier by intertwining both the climate trends using actual climate data as well as future scenarios (described earlier) with climate perceptions and the use of local indigenous knowledge systems and practices (LIKSP). These social aspects of climate are extremely important despite the debate about their use and integration (Dunlap and Brulle 2015; Trumbo and Shanahan 2000; Mubaya 2011). This is because if any awareness campaigning, knowledge sharing, policy formulation, reformulation and implementation as well as intervention programmes are to work effectively then they must involve the participation of communities, which is a whole different ball game from 'looking up the sky and predicting weather and projecting climate'. In addition, some climate impact and vulnerability studies are beginning to emerge where historical, current and projected impacts of changes in climate across several sectors, particularly agriculture, hydrology, health, energy and infrastructural development in SSA are emphasised.

Despite this development, impact and vulnerability assessments (IVA), especially in the SSA, are still scattered, uncoordinated and currently performed at different/ multi-scales such that comparisons across different geographic levels are often not possible (Mubaya et al. 2014). We, however, leave the clutter of IVA and transcend to the zone where even less studies have neither been done nor documented. This is the 'marrying' of climate trends with perceptions and LIKSP. This is because we believe that with such a 'tie-of-the-knot', the IVA as well as interventions become more pliable as the 'marriage' provides impetus and a chance to inform IVA and interventions more faithfully. This then provides a holistic approach and feedback loop that can be used as part of a comprehensive framework for future planning, adaptation, and mitigation programming for a cause (climate change and variability) that has gripped and indeed vexed the entire globe.

For years, local communities have lived and depended on the Zambezi River Basin for the sustenance of their livelihoods and food security, and have relied on indigenous knowledge to predict weather patterns, strategize on coping, and adaptation measures (Manyanhaire 2015; Singini et al. 2015). Indigenous knowledge may be defined as knowledge gained over long periods of observation by communities, and passed on from generation to generation (Boateng 2006). Indigenous people are excellent observers and interpreters of changes to their environment. Their collectively held knowledge that has been accumulated over long periods of time provides invaluable insights into the status of the environment. Indeed, some of the ecological changes noted in this study were provided by observant communities within the basin (see Chapter 3). Different methods and practices have originated within such communities for observing such changes, and these have been passed on from one generation to another. However, although local communities have in the past successfully used their indigenous knowledge to formulate coping and adaptation strategies, the magnitude of changes that are currently taking place has limited their capacity to effectively do so.

Because indigenous knowledge possesses chronological and landscape-specific precision and detail on local climate, it provides ways of observing climate that are different and complementary to 'western' science that normally produces models at a broader scale. The 4th IPCC report acknowledges indigenous knowledge as invaluable for developing sustainable climate change mitigation and adaptation strategies (IPCC 2014b). Combining both indigenous and scientific knowledge on climate change and science-based forecasts helps to produce robust climate forecasting systems that help in the design of effective adaptation strategies (Mubanga and Umar 2014; Soropa et al. 2015). This section provides an overview of climate perceptions on climate change and the use of LIKSP in early warning systems for the communities in the Zambezi River Basin so as to complement the science-based climate analyses systems discussed in the previous sections.

Perceptions of Local People on Climate Change

We conducted a desk study in order to investigate perceptions of local communities within the Zambezi River Basin on climate change and early warning systems. A study by ZVDI (2010), in Kalabo District, Western Province, Zambia which is located within the Upper Zambezi, revealed that although community members did not understand the terms climate change or global warming, they were aware of the effects of seasonal changes in temperature and precipitation patterns. Mubaya et al. (2010 and 2012) concluded that in as much as farmers from Monze and Sinazongwe districts in semi-arid Southern Zambia as well as Lupane and Lower Gweru districts in South-Western Zimbabwe reported changes in local climatic conditions which were consistent with climate variability, there was a disparity in allotting the input of climate variability as well as other factors to detect impacts on the agricultural and their socio-economic system. Mubaya et al. (2012) also highlight that climate variability was the most significant among multiple stressors that aggravate livelihood insecurity for the communities. In a separate study (ZDVI 2010), it was found that the Lozi people's livelihoods and culture are closely linked with the flooding regimes of the Barotse Flood Plain and because of this; climate change poses serious threats to their way of life, their existence and culture. Effects of climate changes and variability that have been noted by communities in this area include the following:

▶ Extreme weather, e.g. excess heat, harsh sun that dry seedlings before they geminate;

▶ Prolonged dry spells;

▶ Reduced precipitation and intermittent severe droughts;

▶ Unexpected changes in seasons;

▶ Delayed onset of rains from October to November (used to be September);

- ▶ Early floods (used to appear in January but now start in December);
- ▶ Extinction of some species (plants, birds, fish);
- ▶ Erosion of culture by, for instance, cancellation of the traditional ceremony of Kuomboka (that means coming out of the water) due to inadequate floods;
- ▶ Severe storms;
- ▶ Strong winds;
- ▶ Floods that arrive at a faster pace, are deeper and result in more damage to infrastructure and loss of crops and food shortages; and
- ▶ Increased health problems including malaria. The IPCC (2014a) expects increased warming in the tropics to expand the malaria zone with increased incidences of malaria expected.

Local Perspectives: A Case of the Manyame Catchment

This case study investigated the perceptions towards climate vis-à-vis availability of natural resources and the resultant coping strategies among river-dependent rural households in St Ruperts (Runene Village), Makonde District, Mashonaland West, Zimbabwe (Box 5.1). The communities are dependent on the Mupfure River which is within the Manyame Catchment, a major tributary of the Zambezi River.

Use of LIKSP for Climate Forecasting

Although numerous studies have been conducted on climate change and variability along the Zambezi basin, for some parts, e.g. the Upper Zambezi, only a few of them focus on the communities' indigenous knowledge about climate change and indigenous indicators for climate forecasts. This section highlights some of the findings from the limited, but available literature. Traditional knowledge systems use indicators which can be largely classified into three broad categories: atmospheric/ meteorological, biological and geographical. Indicators include use of the moon and sun, whilst biological indicators consist of cues such as tree flowering, bird and insect behaviour and movement. Temperature extremes and wind movements make up the geographical indicators (Simelane 2014). Two of the three broad categories are detailed in the following paragraphs.

Box 5.1: Perceptions of climate and natural resources in Mupfure River catchment, Zimbabwe. A contribution by O. Kupika, B. Utete, C. Mapingure and W. Muzari

Community perceptions on availability and benefit of various natural resources in the Mupfure River catchment indicate that 43.6 per cent (n=24) of the women depend on the river for resources whilst men 56.3 per cent (n=31) harvest resources from the river. The community in Makonde perceived the effects of climate variability on natural resources as the disappearance of vleis (Makan'a in Shona), loss of indigenous tree species such as *Brachystegia boehemii* (mupfuti), *Julbernadia globiflora* (Munondo), *Pericopsis angolensis* (Muwanga), *Pterocarpus angolensis* (Mubvamaropa), *Diospyros mespiliformis* (Mushuma), and reduced flow of rivers, particularly the Mupfure River. Responses from key informants and FGDs indicate that the communities mainly depend on the rivers for fishing, gardening, and irrigation.

There is, however, a decrease on river dependence for building, brick molding and sand collection. In response to extreme events such as droughts, the majority of the respondents (75 per cent n= 64) grazed their cattle along river. Respondents' perceptions pointed to a further decline in river flow and volume which could in turn negatively affect livelihood activities such as fishing and gardening. Excessive water abstraction and siltation of the river beds due to a combination of natural factors and anthropogenic factors (artisanal gold panning, stream bank cultivation and sub-catchment deforestation) is also a cause for concern. Communities noted that climate change and variability as well as other non-climatic factors could be contributing towards the decline in river-based goods and services. For instance, changes in river flow due to reduced rainfall amounts could be responsible for other changes in native fish assemblages. Climate change and variability and other non-climatic factors could also be affecting the phenology of indigenous tree species located along Mupfure River. River-based households have also shifted their riparian-based natural resource utilization patterns in response to variations to river flow regimes.

Atmospheric

Essentially, indigenous forecasting is not solely based on personal experience but also on trend analysis (Kolawole et al. 2014; Mapfumo et al. 2015). For example, a preceding bad season signifies a better season to come and vice versa (Orlove et al. 2010). Mapfumo et al. (2015) cite a case of farmers in Makoni District of Zimbabwe who have traced the changes in five precipitation regimes that had for ages indicated the specific stages of precipitation such as the: (i) onset of the winter season at the end of May (Mavhurachando), (ii) rains coming in August after the processing of grains (Gukurahundi), (iii) late September marking the end of wild fires (Bumharutsva), (iv) hastening growth of new tree leaves in October (Bvumiramutondo), and (v) marking the beginning of the rainy season in October/November (Nhuruka). These case studies illuminate on traditional indicators as affected by changes in precipitation patterns to an extent that they may mislead farmers. Farmers rely on these meteorological indicators for farming practices including securing marketing and trade arrangements for food security (Mapfumo et al. 2015).

Biological

These entail cues such as tree flowering, fruiting, leaf bursts, and insect movement and behaviour. Studies in southern Africa for the past decade have started to show shifts in the flowering patterns of trees in association with El Niño events, with indications that if certain trees bear fruit at certain periods of time, there will either be a good or poor precipitation season (Curran et al. 1999). In Zimbabwe, the disappearance and delayed fruiting of trees such as Maroro, Tsambatsi and Hute and, on the other hand, the profuse fruiting of the Muhacha tree, including the delayed regrowth of grasses from August to October have for a long time indicated droughts to come (Mapfumo et al. 2015). However, it is important to understand that the shifting of tree patterns may render this indicator less reliable but concurrently, it is also important to recognize the significance of the indigenous forecasts for planning purposes at this level. While there is evidence of indigenous indicators based on animal behaviour, these indicators are by far the least commonly used of the various kinds of indicators, although the highlights on LIKSP from a river system in Malawi reflect otherwise (Box 3). Sounds from insects that emerge from overwintering/hibernation tend to signal the start of a rainy season and planning by farmers in Zimbabwe (Mapfumo et al. 2015).

In Zambia, Simelane (2014) noted that scientific crafting of climate scenarios largely relies on complicated climate modelling. Therefore, traditional communities only need a little of scientific knowledge to complement their LIKSP of forecasting and adapting to climate change. The same author advocates combining LIKSP with scientific forecasts for improved measures of adaptation to climate change. As projected by scientists, the Zambezi River Basin is expected to experience some floods as an effect of climate change. The people of Zambia already have a traditional knowledge system pertaining to escaping floods. This is in the form of Kuomboka ceremony which largely involves moving from lower grounds to higher grounds to evade flooding effects of the Zambezi River (Simelane 2014); translating into a form of LIKSP which is an adaptation to climate change.

In Malawi, LIKSP indicators mainly focus on the occurrence of floods and droughts (Box 3). While some of the LIKSP can be verified scientifically, most of them have no scientific bearing on the occurrence of floods and droughts. The few that have been verified scientifically include what is presented in Box 5.2. In Tanzania, similar to the rest of the Zambezi River Basin's riparian countries, LIKSP are used to forecast climatic factors such as precipitation and droughts. As noted earlier, there is an increase of unpredictability of rains in recent times that can be tied to climate change. Communities in Tanzania use several indicators to forecast climate trends such as early or delayed onsets of the rainy season, which determine when to prepare their farms and when to plant. The use of plants, insects, birds and animals to predict weather patterns is commonplace in Tanzania, in the southern highlands in particular (Chang'a et al. 2010; Kangalawe et al. 2011; Kijazi et al. 2012). The Mipalamba tree—which flowers during the dry season—is used to forecast droughts or less rains when they produce little flowers; while early sprouting of the Mihango and Mipogoro (*Acacia* sp.) trees indicate early onset of and good rains thereafter respectively (Kangalawe et al. 2011). Flowering of mango trees also indicate either early or late rains depending on when the flowering starts. Birds have also been used as indicators of the onset of rains. For example, when the Kolekya birds start singing, farmers also start preparing their farms because the singing signifies that the rains are close (Chang'a et al. 2010; Kangalawe et al. 2011; Kijazi et al. 2012).

Despite the wealth of local indigenous knowledge in forecasting the local climates, the communities have noticed shifts in and unpredictability of precipitation seasons that are adversely affecting the farming seasons and food security in the region. For most parts of the basin, there is an increasing trend in drought events while temperatures are also increasing. These changes are affecting the different indicator species that are used by the local communities in forecasting precipitation in particular. Drought events are not only forcing communities to shift into irrigated agriculture, but also will affect the entire agricultural productions thereby exacerbating food insecurity and water stress in the region (Liwenga et al. 2009). While LIKSP are vital in informing both the communities and science-based decisions, there is a need to conduct more LIKSP studies in watersheds (Kijazi et al. 2012). This is because, currently, there is very little LIKSP information that can be reliably used for coping and adaptation purposes. Moreover, the fact that traditional indicators of weather are under pressure from changing patterns implies the need for documentation and institutionalization of traditional knowledge and practices in conjunction with modern scientific forecasting.

Box 5.2: Local indigenous knowledge and practices (LIKSP) in Malawi

One of the signs of impending floods is when frogs and hippos start migrating to the river banks and the flood plain, away from the fast flowing river—this behaviour portends severe flooding. Generally, frogs and hippos are able to notice increases in flow velocities of rivers, hence they move away to the banks and the flood plain to avoid being washed away by floods. This observation was manifested during the 2015 floods that wreaked havoc in the Lower Shire Valley as most of the frogs and hippos moved away from the Shire River. People who were able to notice this observation evacuated to dry land in good time thereby avoiding being swept away by the flash floods; The blowing of strong southeast trade winds over an area during the rainy season is a sign that dry spells will be prevalent during the season—this can be proved scientifically since the rainy season in Malawi is characterized by the prevalence of the ITCZ and the dominance of northeast trade winds, which are generally weak and bring rain to the entire country; prolonged blowing of northeast trade winds (mpepoyam poto) is indicative of high precipitation occurrence, and therefore high chances of flooding; and when the nests of 'Chosos' (a type of wild bird) point to the north—this is indicative of the dominance of southeast trade winds during the rainy season and hence little rains will take place during the season resulting in drought.

'Nanzeze' (a local bird) flying very close to the ground—high precipitation/ flooding;

Prolonged sound production by a wild animal locally known as mbiru— increase in water levels in rivers/flooding;

Extremely high temperatures during the rainy season—high precipitation/ flooding;

Early sprouting of trees during the rainy season—high precipitation/ flooding;

Availability of mangoes in large numbers-occurrence of erratic rains/dry spells;

Prevalence of cold weather during the rainy season—erratic rains/dry spells;

High populations of army worms—low precipitation/dry spells;

Too many grasshoppers (mainly 'anunkhadala')—low precipitation/dry spells;

High rate of flowering of mango trees-low precipitation/drought;

► The frequent chirping of birds known as paradise fly catcher (Nanthambwe) in the month of October—high precipitation/flooding

► The flowering of fig-trees (Ficus burkei)—good rainy season;

► The presence of sparrow, egret, and drongo birds—high precipitation/ flooding;

► Abundance of crickets—heavy rains, therefore flooding;

► Tips of anthills pointing south-prevalence of northeast trade winds, therefore heavy rains and flooding.

Source: Simelane (2014)

Discussion

On Climate Analyses

According to the SRES A1B scenario of the GCM data, the mid-century years are projected to be warmer in the ZRB than the base period of the last 30 years (Hawkins and Sutton 2009). This warming was reflected in our study as mean annual temperatures in the basin were projected to range between 2.08°C to

2.9°C above the 1961-1990 average (Figure 5.5c) while the projected increase of basin mean surface temperature by the mid-century period (2045-2065) relative to 1961-1990 is likely to vary from 1.7°C to 2.03°C under A1B (Table 5.5). Our model projections match the historical data trends which showed significant annual warming in maximum temperature in Manyame, Shire and Songwe (Figure 5.2), as well as minimum temperature in Barotse, Shire and Songwe (Figure 5.3). All five catchments in the basin showed increasing trends of mean seasonal minimum and maximum historical temperatures (Table 5.3 and Table 5.4). In addition, our study also reflected the heterogeneity of the sub-catchments under study in that while temperatures have increased in the past thirty (~30+) years, they have done so at differential rates and are projected to likely vary in future across the five sub-catchments.

The Zambezi basin showed spatial variation of precipitation (see Figure 5.6), leading to large variability among the sub-catchments (see Figure 5.6). While Barotse and Ruhuhu showed annual increasing trends in precipitation, Manyame, Shire, and Songwe had a decreasing trend (see Table 5.6). The GCM model datasets analysis also showed general agreement with the historical time series (Figure 5.6). Several coherent anomaly patterns were evident over the historical precipitation pattern (Figure 5.6a). One of the most prominent patterns was the strong negative anomaly in the Shire catchment in Malawi and Barotse catchment in Zambia (see Table 5.6). This is a cause for concern given the countries' dependency on the same water sources along the Zambezi that is from the Shire (Malawi), Barotse and Kafue (Zambia) and Kariba (Zambia/Zimbabwe) not only for water use but for hydroelectricity sources, tourism attractions, navigation as well as proper functioning of ecosystems therein. As shown in Figure 5.5c, precipitation over the Zambezi River Basin is projected to rise above the 1961-1990 average in mid-century in the central and north-eastern regions, whereas the south, south-east and north-west region will remain drier. A similar trend was reported by Willis et al. (2013) who projected that the southern region of the Zambezi basin was to become slightly drier than the northern region by the mid-century. At the same time, less than 300 mm in total precipitation is expected over the south eastern part of the basin in the Shire catchment (Figure 5.7b). In contrast, the Ruhuhu and Songwe catchments which are located north-east of the basin and Barotse and Manyame catchment in the central part of the basin (Figure 5.6) are expected to receive annual precipitation above normal ranges. However, it is important to note that the historical decreased or increased trend of precipitation for all catchments in the basin were insignificant (see Table 5.6).

Our results largely compare well, rather than contrast, with some of the majority of global climate models. For example, the southern African region where the ZRB is mostly situated has been projected to experience increases in aridity in the future with implications for reduction of available water for livelihoods (Odada 2013; DWC 2003; IPCC 2001, 2007, 2014a, 2014b), which agrees well with this study. In contrast, however, the increase in maximum temperature over the period 1980-2014 in our study is lower than reported in other studies (Tauya 2010; Shongwe et al. 2009). The discrepancy in this regard could be attributed to several reasons. Firstly, it could be the difference in the method used to test temperature trends. While this study used a modified Mann-Kendall technique, which accounts for non-normality assumptions, especially poor data quality, Tauya (2010) and Shongwe et al. (2009) used ERA-Interim reanalysis, Taylor diagrams and Pearson's correlation to account for poor data quality. Secondly, the discrepancy between this study and others could be as a result of the differences in the time window of the dataset and therefore different datasets used. Dataset quality is a crucial contributor to model output certainty or uncertainty (Lennard et al. 2015)—reflecting the 'garbage in, garbage out' concept. Similarly, Kling et al. (2014) cite modelling challenges they encountered for the Zambezi as: large basin area, data scarcity and complex hydrology. The lack of consistent, good quality, long-term observational data in Southern Africa has long been singled out as a setback in statistical modelling for climate projections (Kalognomou et al. 2013). The present study used rainfall data from averaged gauging stations that fall within the ZRB catchment and the climate data was updated to 2014. In contrast, other studies used older data sets up to 2010.

In general, the Zambezi Basin region has experienced floods in the north-east and episodes of severe and prolonged droughts in other places. The 2015/2016 rainy season, for instance, has recorded some of the highest temperatures and heat waves ever experienced in the region in comparison to other El Niño years (1998, 2002) with a consequent drought impact in the southern parts of southern Africa and flooding in east Africa (Munich Re Topics Geo 2015). In fact, several countries within the ZRB region, such as Tanzania and Zimbabwe in general, are expected to become water-stressed by 2025 (ZAMCOM et al. 2015). It is also important to note that evaporation in the ZRB is an important climate parameter which determines water balance in the system (see Chapter 7). The basin experiences high evaporation rates (see Table 5.1) while ZAMCOM et al. (2015) report the evaporation of the basin to vary between 1,800 and 2,000 mm and an average of 5 mm per day, and with a potential to reach a maximum 9 mm average per day in hot months such as September and October. Chenje (2000) reports that 65 per cent of all the rain within the basin evaporates as soon as it falls and 20 per cent is lost through evapotranspiration, thus leaving only an average of 15 per cent of the total rainfall as surface runoff. Our results indicated that although temperatures, historic and projected, are rising across the entire basin, the changes and warming rates are different among the sub-basins which necessitates that the sub-catchments should be treated as heterogeneous. This indicates that local climate variability is a strong force in determining localized temperature, particularly in the five river basins under study. This also corresponds well with the heterogeneity in the elevation across the basin (see

Figure 5.1). Elevation has the potential to influence the local temperature: the lower the elevation, the seemingly hotter it becomes and the opposite being true if all things are equal.

Given the spatial differences in climate across the ZRB, it is not surprising that GCM models may miss the regional differences. Uncertainty of GCM prediction at local scale has promoted the use of regional climate models (RCMs) to simulate local climate phenomena. Alternatively, others advocate for GCMs and the coarser resolution, arguing that climate signals are often affected by various processes which occur at larger spatial scales as those covered at a regional level. Therefore, there is a need to validate RCMs in order to understand their usefulness (Giorgi et al. 2009; Kalognomou et al. 2013; Lennard et al. 2015; Pinto et al. 2015; Shongwe et al. 2011; Shongwe et al. 2014; Tadross et al. 2008). In evaluating ten Coordinated Regional Climate Downscaling Experiment (CORDEX) RCMs, Kalognomou et al. (2013) showed that the averaged ensemble outperformed individual models and was able to reproduce seasonal and inter-annual regional climatic features as well as capture the dry (wet) precipitation anomaly associated with El Niño (La Niña) events across the region. However, Kalognomou et al. (2013) provide various instances where individual RCMs were shown to have biases in modelling precipitation. In comparison, Pinto et al. (2015) and Lennard et al. (2015) evaluated several updated CORDEX RCMs over the southern Africa region and found that they depicted the historical seasonal trends of extreme precipitation events well. Despite this, Pinto et al. (2015) cite inadequate and unrepresentative spatial distribution of rain gauges as a setback in capturing credible projections of extreme precipitation events in southern Africa. In comparison, Klutse et al. (2014) and Gbobaniyi et al. (2015) showed that CORDEX models were able to simulate the monsoon events in the west African region. Current research thrusts are that RCMs may be useful in evaluating regional climate but with reservations for local climate.

Risbey et al. (2014) argue that climate models do not replicate recent temperatures; and therefore, should not be relied upon completely for future warming projections (but see Allan et al. 2014 for a different view). Although climate models are known to include all the natural variability scientists know about, they are not designed to predict exact timing of when we will see warming speed up or slow down (Risbey et al. 2014). Understanding natural variability is crucial to predicting how temperatures will change into the future. Model results suggest that climate models can do a good job on those timescales and longer ones provided they capture natural variability well enough (Risbey et al. 2014). The issues of uncertainty when it comes to climate modelling are discussed in the introduction of this chapter. Whilst CORDEX RCMs under the Africa group have been advocated for because they are specific to Africa and the re-ensembles are acceptable (Giorgi et al. 2009; Pinto et al. 2015), the models are still not fully implicit and some biases have been noted (Nikulin et al. 2012; Kalognomou et al. 2013; Pinto et al. 2015). For instance, Kling et al. (2014) show that the two climate models they validated gave different signs for future precipitation change, suggesting large uncertainty. Hence Nikulin et al. (2012) and Pinto et al. (2015) conclusively advocate for the use of multi-model ensemble means as compared to using single models.

Despite the reservations on climate modelling and future projections, we do not, however, advocate for separate national-level analyses of climate when it comes to transboundary systems such as the Zambezi. This is because these man-made boundaries miss the mark and therefore basin countries must make concerted effort towards strengthening both observational data systems and systemic, continuous climate monitoring; improving institutional capacities and collaborated exertions in the holistic evaluation of the ZRB. Our experience in accessing climate data from national meteorological services for research from the four respective countries for the five river basins under study spoke volumes of how far we still are in achieving these ambitious efforts for the basin. The purchase of data and lack of coordination presents hurdles in climate research and development which ideally should then provide research-based evidence to guide policy formulation, implementation and reformulation in the region particularly under the changing climate.

However, a number of coordinated efforts in assessing the ZRB as a whole have been made and some of the organizations that are leading in this are the Zambezi Watercourse Commission (ZAMCOM), Southern African Developing Community (SADC), Southern African Research and Documentation Center (SARDC) and the International Union for Conservation of Nature (IUCN). Some of the outputs from these efforts include the publication of the Zambezi River Basin: Atlas of the Changing Environment as well as the Zambezi Environmental Outlook. The latest copy of the former was in 2012 and the latter was published late in 2015 while the previous version of the same series was last published in the year 2000. This presents a huge gap between the series and often research gaps between when there is no continuous monitoring and evaluation of the Zambezi. Among other changes, the Zambezi Environment Outlook (2015) notes that the basin is characterized by declining water quality, depletion of groundwater and a surge in aquatic invasive species. It also notes that there will be more changes in rainfall patterns in the basin and that a decrease by 10 to 15 per cent in rainfall is expected by 2050. In the present study, this tallies well with only the Shire River basin which had the highest levels of projected precipitation for the mid-century period but not with the rest of the sub-catchments (see Table 5.6). Land and agriculture challenges noted in the Outlook include declining per capita land availability as a result of growing population, soil erosion and fertility decline, land degradation and soil salinization, as well as outbreaks of new strains of diseases. The Outlook also details issues and challenges in the other sectors such as tourism, energy and industrialization.

On LIKSP

Our study showed that Local Indigenous Knowledge and Practices (LIKSP) are commonly used across the Zambezi Basin. In addition, many of these LIKSP are similar: the indicators, which generally did not differ among the different communities, could be broadly classified as meteorological (moon and sun); biological (tree flowering and fruiting; leaf burst; bird and insect movement and behaviour); and geographical (temperature extremes; wind movements). Some bird types are common indicators for short term forecasting (Simelane 2014).

Since LIKSP have been practiced for years and indeed subsume a huge part of the culture of communities in the basin, there is a need to integrate the 'scientific' know-how of climate analyses with LIKSP in climate awareness, interventions and adaptation planning. Often LIKSP have been dismissed as folklore or tales by 'scientists' due to the ever-decreasing capacity of the same to predict weather and climate. However, the physical climate sciences and analyses can strengthen the capacity of LIKSP particularly in this era where unprecedented changes in the climate and its variability are occurring more and more and with the advancement in climate modelling.

In this regard, some players within the ZRB have made some effort to integrate LIKSP with climate sciences. The Zimbabwe Meteorological Services, for instance, working with experts in linguistics, have developed a vernacular dictionary of climate and weather terminology. This came after a realization of the mismatch in climate information sharing and communication particularly in trickling down from the Met services to the local users who are mostly rural based. Similarly, ZVDI (2010) noted that community members in the Kalabo region in western Zambia did not understand climate change or global warming terms, but were sentient of the effects of seasonal changes in precipitation patterns and changes in temperature as did communities in Monze and Sinazongwe (southern Zambia) (Mubaya et al. 2012). Simelane's (2014) comprehensive study of LIKSP in 13 SADC countries advocated for six recommendations which include:

i.   A call for a multi sectoral approach to climate change adaptation;

ii. Sensitization of River Basin Organizations on indigenous community leadership, knowledge and practices;

iii. Documentation and dissemination of local indigenous knowledge and practices for forecasting;

iv. Introduction of indigenous knowledge and practices into the school curriculum;

v.   Institutionalization, Promotion and Commercialization of Indigenous Knowledge and Practices; and,

vi. Protection of ecologically sensitive sites.

In our case, we focused on points i), iii) and vi) of Simelane's recommendations mostly because these points fit aptly within the framework of our research and this book. We consider the ZRB as ecologically sensitive (see Chapter 3) while this chapter focuses on climate dynamics and the hydrology of the ZRB (water resources, access, and governance) are discussed in depth in Chapters 6 and 7. In this regard, we consider the ZRB as not only ecologically sensitive, but its protection crucial. In the Africa Environmental Outlook's Freshwater Resources chapter, ZAMCOM et al. (2015) list several challenges faced in realizing opportunities for freshwater resources in Africa with the ZRB being no exception to them. These include, among others: water governance, tourism, policy implementation and enforcement, information generation and management, knowledge gaps, food security, public health, safe drinking water and sanitation, environmental degradation and financial resources, which are also discussed by Pietersen and Beekman (2006). The last point (finances) as well as gender and youth, climate change, pollution, transboundary issues, IKS and technological developments have been identified as cross-cutting (overarching) issues of concern. Similarly, we also identified LIKSP as cross-cutting, while gender roles were shown to be important in formulating perceptions on natural resources management. Indeed, gender dynamics appear as a strong determinant of perceptions and behaviour when it comes to natural resources management (NRM), climate change and economic considerations, especially with poverty identified as one of the drivers of ecological changes (see Figure 2.7, Chapter 1 and Chapter 2). The gender interface was reflected in the understudied Raffingora community within the Manyame River basin (see Box 1 of Chapter 3).

Women play major roles in NRM and land and water use is largely established by the work of women (UBOS 2006; UWASNET 2009; Mubaya et al. 2012; Asaba et al. 2013; Casarotto and Kappel 2013; Mutopo 2013). The Zambezi Environmental Outlook (ZAMCOM et al. 2015) identifies some of these roles as water collection, irrigation, domestic water use decisions, livestock water supplements and feeding, firewood collection and tilling and tending of farmland within the ZRB. It is also interesting to note that there are more women in the ZRB than men and the majority of these women are rural based and also constitute an important tier of the population as the major sources of labour (Murisa 2009). If any solutions are proposed for the development of the basin especially for NRM, then gender equity and inclusivity must be addressed therein.

Conclusion

It is evident that the climate in the Zambezi River Basin has changed, and is projected to continue changing. Air temperatures have increased and our results indicate that minimum air temperatures are actually warming faster than the maximum temperatures. Apart from just showing catchments with significant decrease or increase trends and of both seasonal and annual precipitation and temperature, this study has also illustrated how historical and future changes of climate may vary across the geographical scale of the Zambezi River Basin. Adding a spatial dimension in this study helped to understand how precipitation and temperature varies in space and time. Both of these dimensions show heterogeneity across the Zambezi. In addition, the frequency of extreme events such as floods and droughts is increasing in the basin. While the natural resources are declining, land is becoming less productive and water becoming an even more scarce resource as a result of the ever-increasing demand for it by the ever-growing human population (see Chapter 3). Therefore, as Kusangaya et al. (2014) reported, the added effects of climate change and warming are certainly an exacerbating stressor, among others.

The communities in the basin use similar weather forecasting indicators — such as animals, plants, insects, temperatures, and winds (Chang'a et al. 2010; Kangalawe et al. 2011; Kijazi et al. 2012) to predict rainfall and droughts. Some of these indicators have indeed been found to serve as early warning systems for droughts and floods in the basin despite the increasing mismatch due to unprecedented climate changes and variability. There is therefore a need to integrate LIKSP in weather forecasting as LIKSP can be understood and followed easily by respective communities (Simelane 2014). We recommend that future studies integrate both inferential and spatial statistics when examining changes and variability in climate across the Zambezi River Basin. This means including and applying LIKSP together with the spatial variation of the climate to support management and decision making.

References

Abbott, L. K. & Murphy, D. V., 2007, 'What is soil biological fertility?', in L.K. & D. V. Murphy, (eds), Soil Biological Fertility — A key to sustainable land use in agriculture, Netherlands: Springer.

Agrawal, A. & Gopal, K., 2013, 'Biomass production in food chain and its role at trophic levels', in A. Agrawal, & K. Gopal, eds, Biomonitoring of Water and Waste Water, India: Springer.

Ahlström, A., Smith B., Lindström, J., Rummukainen, M., & Uvo, C. B., 2013, 'GCM characteristics explain the majority of uncertainty in projected 21st Century Terrestrial Ecosystem Carbon Balance', Biogeosciences, 10, 1517-1528.

Allan, R. P., Liu, C., Loeb, N. G., Palmer, M. D., Roberts, M., Smith, D. & Vidale, P. L., 2014, 'Changes in global net radiative imbalance 1985-2012', Geophysical Research Letters, 41, 5588-5597.

Arnell, N. W., 1999, 'Climate change and global water resources', Global Environmental Change, 9, S31-S50.

Arnell, N. W. 2004. Climate change and global water resources: SRES emissions and socio-economic scenarios. Global Environmental Change, 14, 31-52.

Asaba, R.B., Fagan, G., Kabonesa, C. & Mugumya, F., 2013, 'Beyond distance and time: Gender and the burden of water collection in rural Uganda', *Journal of Gender and Water*, 2, 31-38.

Beilfuss, R., 2012, A risky climate for southern Africa hydro. Assessing Hydrological Risks and Consequences for Zambezi River Basin Dams. Berkeley, CA, USA, *International Rivers*, 56 pp.

Boateng, W., 2006, 'Knowledge management working tool for agricultural extension: The Case Of Ghana', *Knowledge Management For Development Journal*, 2, 19-29.

Bourgeois-Pichat, J., 2008, 'Problems of population size, growth and distribution in Africa', in G. Wolstenholme, & M. O'connor, eds, Man and Africa, London, UK: J. and L. Churchill John Wiley & Sons Ltd.

Bureau, D. P. & Hua, K., 2010, 'Towards effective nutritional management of waste outputs in aquaculture, with particular reference to salmonid aquaculture operations', *Aquaculture Research*, 41, 777-792.

Caesar, J., Janes, T., Lindsay, A. & Bhaskaran, B., 2015, 'Temperature and precipitation projections over Bangladesh and the upstream Ganges, Brahmaputra and Meghna systems', *Environmental Science Process Impacts*, 17, 1047-56.

Cassaroto, C. & Kappel, R., 2013, 'A half empty basket: Women's role in the governance of water resources in Zambia', Management Team, *The Journal of Gender and Water*, 3, 26 pp

Chang'a, L. B., Yanda, P. Z. & Ngana, J., 2010, 'Indigenous knowledge in seasonal rainfall prediction in Tanzania: A case of the South-western Highland of Tanzania', *Journal of Geography and Regional Planning*, 3, 66-72.

Chenje, M., 2000, 'State of the environment: Zambezi River Basin 2000', Harare, Zimbabwe: IUCN/SADC.

Conway, 2009, 'The science of climate change in Africa: impacts and adaptation', Grantham Institute for Climate Change Discussion Paper No 1, Imperial College, London.

Curran, L. M., Caniago, I., Paoli, G. D., Astianti, D., Kusneti, M., Leighton, M., Nirarita, C. E. & Haeruman, H., 1999, 'Impact of El Nino and logging on canopy tree recruitment in Borneo', *Science*, 286, 2184-2188.

Curry, and Webster, 2011, 'Climate science and the uncertainty monster', *American Meteorological Society*, 92, 1667-1682.

Desanker, V. & Justice, C. O., 2001, 'Africa And Global Climate Change: Critical issues and suggestions for further research and integrated assessment modeling', *Climate Research*, 17, 93-103.

Desanker, P. V. & Magadza, C., 2001, Africa. Chapter 10 of the IPCC Working Group II, Third Assessment Report, Cambridge, UK: Cambridge University Press.

Dugan, P., 1993, 'Managing the wetlands. People and rivers: Africa', *People Planet*, 2, 30-3.

Dunlap, R. E. & Brulle, R. J., eds, 2015, Climate change and society: sociological perspectives, Oxford, UK: Oxford University Press.

DWC, 2003, 'Climate changes the water rules-How water managers can cope withtoday's climate variability and tomorrow's climate change', in P. Kabat, & H. van Schaik, 'Dialogue on Water and Climate', 3rd World Water Forum Report, Delft. ISBN-90-327-0321-8.

Easterling, W. & Apps, M., 2005, 'Assessing the Consequences of Climate Change for Food and forest Resources: A View from the IPCC', *Climatic Change*, 70, 165-189.

FAO, 2014, 'Agriculture "engine of growth" that Africa needs' / FAO Director-General marks launch of the African Year of Agriculture and Food Security and 2025 zero hunger target. African Press Organisation. Database of Press Releases Related to Africa, 01/29/2014.

Gaitan, J. J., Oliva, G. E., Bran, D. E., Maestre, F. T., Aguiar, M. R., Jobbagy, E. G., Buono, G. G., Ferrante, D., Nakamatsu, V. B. & Ciari, G., 2014, 'Vegetation structure is as important as climate for explaining ecosystem function across Patagonian rangelands', *Journal of Ecology*, 102, 1419-1428.

Gbobaniyi, E., Sarr, A., Sylla, M.B., Diallo, I., Lennard, C., Dosio, A. et al., 2014, 'Climatology, annual cycle and interannual variability of precipitation and temperature in CORDEX simulations over West Africa', *International Journal of Climatology*, 34, 2241-2257.

Giorgi, F., Jones, C. And Asrar, G.R., 2009, 'Addressing climate information needs at the regional level: the CORDEX framework', World Meteorological Organization (WMO), *Bulletin*, 58, 175-183.

Gitay, H., Brown, S., Easterling, W., Jallow, B. & Al, E., 2001, 'Chapter 5. Ecosystems and Their Goods and Services', in J. J. Mccarthy, O. F. Canziani, N. A. Leary, D. J. Dokken, & K. S. White, eds, Climate Change 2001: Impacts, Adaptations, and Vulnerability. Contribution of Working Group II to the Third Assessment Report of the International Panel on Climate Change. IPCC/Cambridge University Publication Press.

Hawkins, E. & Sutton, R., 2009, 'The Potential To Narrow Uncertainty', in Regional Climate Predictions. *Bulletin of the American Meteorological Society*, 90, 1095-1107.

Hendrix, C. S. & Glaser, S. M., 2007, 'Trends and triggers: Climate, climate change and civil conflict in Sub-Saharan Africa', *Political Geography*, 26, 695-715.

Henson, R., 2011, 'National Academics Report', in R. Henson, ed., Warming impacts by degree', Viewed on 7 December 2015, at www.dels.nas.edu/warming_world_final.

Hulme, M., Doherty, R., Ngara, T., New, M. & Lister, D., 2001, 'African climate change: 1900-2100', *Climate Research*, 17, 145-168.

Hulme, P. E., 2005, 'Adapting to climate change: Is there scope for ecological management in the face of a global threat?' *Journal of Applied Ecology*, 42, 784-794.

IPCC, 2001, 'Climate change 2001: Impacts, adaptation and vulnerability', in: J. J. Mccarthy, O. F. Canziani, N. A. Leary, D. J. Dokken, & K.S. White, eds, Contribution of Working Group II to the third assessment report of the Intergovernmental Panel on Climate Change. Cambridge, UK: Cambridge University Press.

IPCC, 2007, 'Climate Change 2007: Chapter 4: Hydrology and water resources', in Parry, O. F. Canziani, P. J Palutikof, Van Der Linden & C. E. Hanson, eds, Impacts, Adaptation And Vulnerability: Contribution Of Working Group II to the forth assessment report of the Intergovernmental Panel on Climate Change. Cambridge, UK: Cambridge University Press.

IPCC, 2014a, Climate Change 2014, Impacts, Adaptation, and Vulnerability. Part A: Global and Sectoral Aspects. Contribution of Working Group II to the Fifth Assessment Report of the Intergovernmental Panel on Climate Change. Cambridge, United Kingdom and New York, NY, USA, Cambridge University.

IPCC, 2014b, 'Climate Change 2014: Synthesis Report. Contribution of Working Groups I, II and III to the Fifth Assessment Report of the Intergovernmental Panel on Climate Change', in R. K. Pachauri, & L. A. Meyer, (eds, Geneva, Switzerland: IPCC.

IPCC, 2014c, 'Climate Change; Reports Outline Climate Change Findings from C.D. Idso and Colleagues (S. Fred Singer And The Nongovernmental International Panel On Climate Change)', The Business of Global Warming, Cambridge, UK: Cambridge University Press, 506 pp.

Jayaram, K., Riese, J. & Sanghvi, S., 2010, Africa's path to growth: Sector by sector-agriculture; Abundant opportunities, McKinsey and Company, Chicago, Munich.

Kalognomou, E-A., Lennard, C., Shongwe, M., Pinto, I., Favre, A., Kent, M., Hewitson, B., Dosio, A., Nikulin, G., Panitz, H-J. And Büchner, M., 2013, 'A Diagnostic Evaluation of Precipitation in CORDEX Models over Southern Africa', *Journal of Climate*, 26, 9477-9506.

Kangalawe, R., Mwakalila, S. & Masolwa, P., 2011, 'Climate Change Impacts, Local Knowledge and Coping Strategies in the Great Ruaha River Catchment Area, Tanzania', *Natural Resources*, 02, 212-223.

Kijazi, A. L., Chang'a, L. B., Liwenga, E. T., Kanemba, A. & Nindi, S. J., 2012.' The use of indigenous knowledge in weather and climate prediction in Mahenge and Ismani Wards, Tanzania', pp 42-48. Proceedings of the first Climate Change Impacts, Mitigation and Adaptation Programme Scientific Conference, 2012.

Kling, H., Stanzel, P. & Preishuber, M., 2014, 'Impact modelling of water resources development and climate scenarios on Zambezi River discharge', *Journal of Hydrology: Regional Studies*, 1, 17-43.

Klutse, N. A. B., Sylla, M. B., Diallo, I., Sarr, A., Dosio, A., Diedhiou, A. et al. 2015, 'Daily characteristics of west African summer monsoon precipitation in CORDEX simulations', *Theoretical and Applied Climatology*, DOI: 10.1007/s00704-014-1352-3.

Kolawole, O. D., Wolski, P., Ngwenya, B. & Mmopelwa, G., 2014, 'Ethno-meteorology and scientific weather forecasting: small farmers and scientists' perspectives on climate variability in the Okavango Delta, Botswana', *Climate Risk Management*, 4, 43-58.

Kupika, O. L., Utete, B., Chibememe, G., Danha, C., Mapingure, C. & Muzari, W., 2012, 'Climate change and variability as perceived by river dependent rural communities in Makonde District, Chinhoyi, Zimbabwe', *Zimbabwe Journal of Technological Sciences*, 1, 35-46.

Kusangaya, S., Warburton, M. L., Van Garderen, E. A. & Jewitt, G. P. W., 2014, 'Impacts of climate change on water resources in southern Africa: A review', *Physics and Chemistry of the Earth*, 67-69, 47-54.

Lennard, C., Dosio, A., Nikulin, G., Pinto, I. And Seid, H., 2015, 'Quantifying uncertainty in observational rainfall datasets', in: EGU General Assembly Conference Abstracts, 17, 901 pp.

Li, Z., Zhang, Y., Yu, D., Zhang, N., Lin, J., Zhang, J., Tang, J., Wang, J. & Mu, C., 2014, 'The influence of precipitation regimes and elevated CO_2 on photosynthesis and biomass accumulation and partitioning in seedlings of the rhizomatous perennial grass (*Leymus chinensis*), PLoS ONE, 9, 1-9.

Liwenga, E. T. & Kangalawe, R. Y. M., 2009, Climate Change/ Variability and Implications on Agricultural Production and Livelihoods in the Southern Highlands of Tanzania', in P. S. Maro, & A. E. Majule, eds, Strengthening local agricultural innovations to adapt to climate change in Botswana, Malawi, South Africa and Tanzania. SADC Secretariat, Dar es Salaam, pp. 124-135.

Magadza, C. H. D., 2010, 'Indicators of above normal rates of climate change in the Middle Zambezi Valley, Southern Africa', Lakes and Reservoirs: Research and Management, 15, 167-192.

Magadza, C. H. D., 2011, 'Indications of the effects of climate change on the pelagic fishery of Lake Kariba, Zambia-Zimbabwe', *Lakes and Reservoirs: Research and Management*, 16, 15-22.

Manyanhaire, I.O., 2015, 'Integrating indigenous knowledge systems into climate change interpretation: perspectives relevant to Zimbabwe', *Greener Journal of Educational Research*, 5, 27-36.

Mapfumo, P., Mtambanengwe, F. & Chikowo, R., 2015, 'Building on indigenous knowledge to strengthen the capacity of smallholder farming communities to adapt to climate change and variability in southern Africa', *Climate and Development*, 1-11.

Mendelsohn, R., Dinar, A. & Dalfelt, A., 2000, 'Climate change impacts on African agriculture. Preliminary analysis prepared for the World Bank, Washington, District of Columbia, 25 pp.

Millard, S. P., 2013, EnvStats: An R Package for Environmental Statistics, Springer, New York.

Mopelwa, G. & Blignaut, J., 2014, 'The Okavango Delta: the value of tourism', *South African Journal of Economic and Management Sciences*, 9, 113-127.

Mubanga, K.H. & Umar, B.B., 2014, 'Climate Variability and Change in Southern Zambia: 1910 to 2009', in 2014 International Conference on Intelligent Agriculture (ICOIA), 94-100 pp.

Mubaya, C. P., 2006, 'CIRAD-CASS Report', in CIRAD-CASS, ed., Understanding policy and legislative policies for Natural resource management: An assessment of governance, uses and livelihood options surrounding selected high value resources by small scale communities in Africa (The case of Omay communal lands, Zimbabwe, Cirad-cass), 5, 170-186.

Mubaya, C. P., 2011, 'Carving a niche for the social sciences in inter-disciplinary research on climate change adaptation and agriculture in southern Africa', Paper read at the ISSC-CIPSH General Assembly Joint Symposium, 01 May 2011, Nagoya.

Mubaya, C. P., Njuki, J., Liwenga, E., Mutsvangwa, E. P. & Mugabe, F. T., 2010, 'Perceived Impacts of Climate Related Parameters on Smallholder Farmers in Zambia and Zimbabwe', *Journal of Sustainable Development in Africa*, 12(5), 170-186.

Mubaya, C. P, Njuki, J., Mutsvangwa, E. P., Mugabe, F. and Nanja, D. H., 2012. 'Climate variability and change or multiple stressors? Farmer perceptions regarding threats to livelihoods in Zimbabwe and Zambia', *Journal of Environmental Management*, 102, 9-17.

Munich Re-Topics Geo, 2015.Natural catastrophes 2015: analyses, assessments, positions, 2016 Issue. https://www.munichre.com/site/.../08875_Topics_Geo_2015_en.pdf. Viewed on 16 June 2016.

Murisa, T., 2009, 'An analysis of emerging forms of social organisation and agency in the newly resettled areas, the case study of Goromonzi and Zvimba Districts', Thesis (PhD), Rhodes University, Grahamstown.

Mutopo, P., 2013, 'The granary is never empty': Female agency, land access and livelihood security in the face of non-permanent mobility after fast track land reform in Mwenezi District, Zimbabwe, Brill and African Studies Centre, Leiden Academic Publishers, Leiden, Netherlands.

Mwingira, C. E., Pallangyo, M. E., Felix, R., Pima, N. & Meing'ataki, S. S., 2011, 'Impacts of climate change on biodiversity and community livelihoods in the Katavi ecosystem, Externship Report: Tanzania. Education and Training Program on climate change and biodiversity in the Albertine Rift', The International START Secretariat, Washington DC.

Ndebele-Murisa, M. R., Mashonjowa, E. & Hill, T., 2011, 'The implications of a changing climate on the Kapenta fish stocks of Lake Kariba, Zimbabwe', *Transactions of the Royal Society of South Africa*, 66, 105-119.

Ndebele-Murisa, M.R., Hill, T. & Ramsay, L., 2013, 'Validity of downscaled climate models and the implications of possible future climate change for Lake Kariba's Kapenta fishery', *Environmental Development*, 5, 109-130.

Ndhlovu, A., 2013, 'Changing environment in the Zambezi River Basin', *The Zambezi*, 8, 1-8.

Nikulin, G., Jones, C., Giorgi, F., Asrar, G., Büchner, M., Cerezo-Mota, R., Christensen, O.B., Déqué, M., Fernandez, J., Hänsler, A. & Van Meijgaard, E., 2012, 'Precipitation climatology in an ensemble of CORDEX-Africa regional climate simulations', *Journal of Climate*, 25, 6057-6078.

Odada, E. O., 2013, 'Our Freshwater Under Threat-Vulnerability of Water Resources to Environmental Change in Africa', Nairobi, Kenya: University of Nairobi.

Ohlberger, J., Mehner, T., Staaks, G. & Hölker, F., 2012, 'Intra-specific temperature dependence of the scaling of metabolic rate with body mass in fishes and its ecological implications', *Oikos*, 121, 245-251.

Orlove, B. & Caton, S. C., 2010, Water sustainability: Anthropological approaches and prospects', *Annual Review of Anthropology*, 39, 401-415.

O'Reilly, C.M., Alin, S.R., Plisnier, P.D., Cohen, A.S. & Mckee, B.A., 2003, 'Climate change decreases aquatic ecosystem productivity of Lake Tanganyika, Africa', *Nature*, 424, 766-768.

Pietersen, K. & Beekman, H., 2006, 'Freshwater', in UNEP, ed., Africa Environment Outlook 2: Our environment, our wealth. Nairobi, Kenya: UNEP.

Pinto, I., Lennard, C., Tadross, M., Hewitson, B., Dosio, A., Nikulin, G., Panitz, H.J. and Shongwe, M.E., 2015, 'Evaluation and projections of extreme precipitation over southern Africa from two CORDEX models', *Climatic Change*, 1-14.

Pricope, N. G. & Binford, M. W., 2012, 'A spatio-temporal analysis of fire recurrence and extent for semi-arid savanna ecosystems in Southern Africa using moderate-resolution satellite imagery', *Journal of Environmental Management*, 100, 72-85.

Ringler, C., 2010, 'Climate Change and Hunger: Africa's Smallholder Farmers Struggle to Adapt Changement climatique et famine: les petits exploitants africains peinent à s'ajuster Klimawandel und Hunger: Kleinbauern in Afrika haben Schwierigkeiten bei der Anpassung', *EuroChoices*, 9, 16-21.

Risbey, J. S., Lewandowsky, S., Langlais, C., Monselesan, D. P., O/'Kane, T. J. & Oreskes, N., 2014, Well-estimated global surface warming in climate projections selected for ENSO phase', *Nature Climate Change*, 4, 835-840.

SADC, 2007, Annual Report 2006-2007. Gaborone, Botswana: Southern African Development Community (SADC), Gabarone, Botswana.

Saghir, J., 2014, 'Global challenges in agriculture and the World Bank's response in Africa' *Food and Energy Security*, 3, 61-68.

Schaeffer, M., Baarsch, F., Balo, G., De Bruin, K., Calland, R., Fallasch, F. Melkie, M. E. 2015, 'Africa's Adaptation Gap Report' in M. Schaeffer, F. Baarsch, R. Munang, & C. Baxter, eds, 'Bridging the gap-mobilising resources', AMCEN, UNEP, Climate Analytics & African Climate Finance Hub.

Schefuss, E., Kuhlmann, H., Mollenhauer, G., Prange, M. & Patzold, J., 2011, 'Forcing of wet phases in southeast Africa over the past 17,000 years', *Nature*, 480, 509-12.

Schmidhuber, J. & Tubiello, F. N., 2007, 'Global food security under climate change', *Proceedings of the National Academy of Sciences*, 104, 19703-19708.

Scott, E. M. & Gemmell, J. C., 2012, 'Spatial statistics-A watery business', *Spatial Statistics*, 1, 121-132.

Shongwe, M. E., Lennard, C., Liebmann, B., Kalognomou, E., Ntsangwane, L. & Pinto, I., 2014, 'An evaluation of CORDEX regional climate models in simulating precipitation over Southern Africa', *Atmospheric Science Letters*, 16, 199-207.

Shongwe, M. E., Van Oldenborgh, G. J., Van Den Hurk, B. J. J. M., De Boer, B., Coelho, C. A. S. & Van Aalst, M. K., 2009, 'Projected Changes in Mean and Extreme Precipitation in Africa under Global Warming. Part I: Southern Africa', *Journal of Climate*, 22, 3819-3837.

Shongwe, M.E., Van Oldenborgh, G.J., Van Den Hurk, B. And Van Aalst, M., 2011, 'Projected changes in mean and extreme precipitation in Africa under global warming. Part II: East Africa', *Journal of Climate*, 24, 3718-3733.

Shurin, J. B., Clasen, J. L., Greig, H. S., Kratina, P. & Thompson, P. L., 2012, 'Warming shifts top-down and bottom-up control of pond food web structure and function', *Philosophical Transactions of the Royal Society B: Biological Sciences*, 367, 3008-3017.

Simelane, Q. G., 2014, 'Local Indigenous Knowledge Systems and Practices (LIKSP) and how they contribute to enhancing climate resilience of communities in the SADC region', *GWPSA Water Dialogue*, 2014, 8-11 pp.

Singini, W., Tembo, M. & Banda, C., 2015, 'Forecasting Climate Change Patterns For Bolero Extension Area in Malawi', *Journal of Climatology and Weather Forecasting*, 3, 1-5.

Soropa, G., Gwatibaya, S., Musiyiwa, K., Rusere, F., Mavima, G.A. & Kasasa, P., 2015, 'Indigenous Knowledge System Weather Forecasts As A Climate change adaptation strategy in smallholder farming systems of Zimbabwe: Case study of Murehwa, Tsholotsho and Chiredzi districts', *African Journal of Agricultural Research*, 10, 1067-1075.

Tadross, M., Randriamarolaza, L., Rabefitia, Z. & Zheng, K. Y., 2008, 'Climate change in Madagascar; recent past and future', Washington, D.C: World Bank.

Tauya, E., 2010, 'Responding to climate change impacts: adaptation and mitigation strategies as practised in the Zambezi River Basin', Harare, Zimbabwe: Southern African Research and Documentation Centre.

Trumbo, C. W. & Shanahan, J., 2000, 'Social research on climate change: where we have been, where we are, and where we might go', *Public understanding of Science*, 9, 199-204.

UBOS, Uganda Bureau of Statistics, 2006, 2006 *Statistical Abstract*. Kampala, Uganda: UBOS.

Unganai, L. S., 1996, 'Historic and future climatic change in Zimbabwe', *Climate Research*, 6, 137-145.

Uwasnet, 2009, 'Report for the Financial Year 2008/9-NGO Group Performance in the Ugandan Water and Sanitation Sector', Kampala, Uganda: Uganda Water and Sanitation Network (UWASNET).

Wallenstein, M. D. & Hall, E. K., 2012, 'A trait-based framework for predicting when and where microbial adaptation to climate change will affect ecosystem functioning', *Biogeochemistry*, 109, 35-47.

Weyl, O. L. F., Ribbink, A. J. & Tweddle, D., 2010, 'Lake Malawi: fishes, fisheries, biodiversity, health and habitat', *Aquatic Ecosystem Health & Management*, 13, 241-254.

Willis, K. J., Bennett, K. D., Burrough, S. L., Macias-Fauria, M. & Tovar, C., 2013, 'Determining the response of African biota to climate change: using the past to model the future', *Philosophical Transactions of the Royal Society B: Biological Sciences*, 368, 2012-2049.

World Bank, 2012, 'World Bank-Africa: New aid', *Africa Research Bulletin: Economic, Financial and Technical Series*, 49, 19551C-19552C.

World Bank, 2015, 'Africa region-Second phase of the Lake Victoria environmental management project: additional financing', Washington, D.C: World Bank.

Wrona, F. J., Prowse, T. D., Reist, J D., Hobbie, J. E., Lévesque, L. M. & Vincent, W. F., 2006, 'Climate change effects on aquatic biota, ecosystem structure and function', *AMBIO: A Journal of the Human Environment*. 35, 359-369.

Zakaria, I., Amidu & Maharjan, K., 2014, 'Climate change impact on revenue of major food crops in Ghana: structural ricardian cross-sectional analysis', in K. L. Maharjan, ed., Communities and Livelihood Strategies in Developing Countries. Japan, Springer.

ZAMCOM, SADC, SARDC, 2015, 'Zambezi Environment Outlook 2015', Harare, Gaborone: ZAMCOM, SADC, SARDC.

ZVDI, 2010, 'Second Draft Report', in Lyambai Vulnerability And Adaptation Project Stage 2, ed., 'Community based adaptation to climate change (CBA)', Mongu, Zambia: ZVDI.

6

Hydrology of the Zambezi River Basin

George V. Lugomela, Geoffrey Chavula and Tongayi Mwedzi

Introduction

The Zambezi River Basin (ZRB) is the largest river basin in the Southern African Development Community (SADC) region and lies between 9 to 20° south of the Equator and 18 to 36° east of the Greenwich Meridian (McCartney et al. 2013). The mainstream (the Zambezi river itself) has a total length of about 3,000 km (McCartney et al. 2013). The Basin covers an area of about 1.34 million km² spanning over eight riparian countries of southern Africa, namely Angola, Botswana, Namibia, Malawi, Mozambique, Tanzania, Zambia and Zimbabwe. Only small areas of Tanzania, Botswana and Namibia lie in the Basin whereas Malawi is almost entirely subsumed within the Basin; a larger part of Zambia, more than half of Zimbabwe and significant parts of Mozambique and Angola comprise the Basin (Euroconsult Mott MacDonald 2008; McCartney et al. 2013). Table 6.1 shows the riparian countries' percentage areas in the Zambezi River Basin.

The Zambezi River arises from Kahene Hills at an altitude of 1,500 m in the Upper Zambezi Sub-basin in Zambia before it enters Angola (McCartney et al. 2013). The sub-basin consists mainly of alluvial deposits of Karoo sand which are very permeable resulting in insignificant surface runoff. Figure 6.1 shows the main river network, elevation and the 13 main sub-basins of the Zambezi Basin namely: Upper Zambezi, Lungue-Bungo, Kabompo, Luanginga, Barotse, Cuando-Chobe, Kariba, Kafue, Luangwa, Tete, Mupata, Shire River and Lake Nyasa/Malawi and Zambezi Delta (ZAMCOM et al. 2015). Details of the geology and geomorphology of some sub catchments within the basin are presented in Chapter 2 of this book. The Zambezi River re-enters Zambia at Chavuma Falls and then flows southward to its confluence with the Lungue, Bungo and Kabompo rivers. Shortly downstream of the confluence, it enters the Barotse Sub-basin and later flows into the Barotse wetland. After exiting the wetland, it confluences again with rivers originating in the Cuando-Chobe Sub-basin before entering the Kariba Sub-basin where it flows though the famous Victoria Falls. The Zambezi then flows into Kariba Reservoir (180,000 Mm³) after which it flows through Mupata Sub-basin. Kafue River joins the Zambezi in Mupata Sub basin and Luangwa River joins before it flows into Cabora Bassa Reservoir (51,750 Mm³) which is located in the Tete Sub-basin. From Cabora Bassa, the Zambezi is joined by the Shire River which drains the Lake Nyasa/ Malawi Sub-basin before ending its roughly 3,000 km long stretch into the Indian Ocean through the Zambezi Delta Sub basin.

Table 6.1: Percentage of Riparian Country Areas in the Zambezi River Basin

Country	Zambezi River Basin	
	Area (km²)	Percentage
Angola	1,246,700	11.6
Namibia	824,269	2.9
Botswana	582,730	14.4
Zambia	752,614	71.7
Zimbabwe	390,759	64.3
Malawi	118,484	93.2
Tanzania	945,087	2.9
Mozambique	799,380	17.5

Source: World Bank (2006)

Most of the Zambezi Basin is situated on high plateau of the ancient continent of Gondwana, with elevations ranging from 800 to 1450 masl (World Bank 2010). The major part of the basin lies between 1,000 and 1,300 m, with only a small portion falling below 100 m and above 1,500 m (Figure 6.1). The nature of the topography exhibited by the Zambezi contributes significantly to the high hydropower potential present in the basin. The basin lies on Pre-Cambrian crystalline and metamorphic rocks which are part of the African and Post-African tertiary planation surfaces. The basement aquifers which develop within the weathered regolith and fractured bedrock play an important role in the hydrology of the basin. Low-lying areas of the basin are covered by sediment layers of varying thickness. The top soil is generally shallow and prone to erosion by water and winds in some parts of the basin (McCartney et al. 2013). Gerrits (2005) identified the main soil types as arenosols, greysols and ferralsols which dominate the Upper Zambezi. Other dominant soils are the luvisols that mainly cover the Luangwa and Zambezi Delta sub-basins.

This chapter focuses on the hydrology of the Zambezi River Basin by examining trends in river discharges over time. Trends in discharges, particularly negative ones, have direct impacts on availability of water for different social economic uses as well as ecosystem health, thus affecting provision of ecosystem goods and services, which is the central theme of this book.

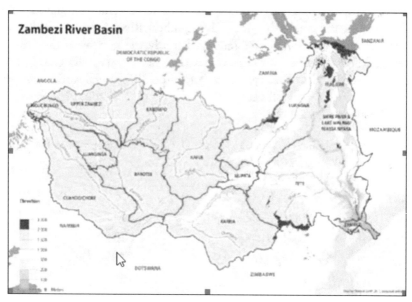

Figure 6.1: The main river network, elevation and the sub-basins of the Zambezi River Basin

Hydrology of the Zambezi River Basin

The seasonal climatic variation in the Zambezi Basin, as in many African regions south of the Sahara, is mainly controlled by the movement of the Inter-Tropical Convergence Zone (ITCZ). The ITCZ is a broad zone in the equatorial low-pressure belt towards which the north-easterly and south-easterly trade winds of the two hemispheres converge. It oscillates randomly across the basin during the rainy season and produces widespread rainfall. The sub-tropical ridge controls the weather over the basin during the dry season which spans over the period May to September. The associated southeast trade winds are stable and there is consequently very little precipitation. However, from October to April, the weather is controlled by the Equatorial trough which moves from the Equator to latitude 20 degrees and back again, popularly known as the ITCZ. This system brings in unstable air which causes convectional rainfall.

The mean annual rainfall at different sub-basins of the Zambezi Basin is shown in Table 6.2. Rainfall variation is very significant throughout the basin occurring mainly between the months of October and April. With varying rainfall averages across the catchment (Trolldalen 1996), the Zambezi River's distribution and occurrence of water resources relies solely on its rainfall, hydrological and evaporation processes (Shela 2000a). The northern parts receive higher amount of rainfall up to 1,400 mm/year or more in the upper reaches and around Lake Nyasa/Malawi while the southern parts (south and south-western) receives slightly less annual rainfall, receiving around 600 mm/year (World Bank 2010; Beilfuss 2012; ZAMCOM et al. 2015). Some recent studies done on rainfall climatology in the countries within the Zambezi River Basin show that while annual amounts of rainfall have remained nearly the same over the years, the onset, end and duration of the rainfall season has dramatically changed, with the rains coming much later than October, and ending much earlier than April, thereby reducing the rainfall season (Beilfuss 2012; IPCC 2007; Nicholson et al. 2013; Shongwe et al. 2009).

Table 6.2: Mean Annual Precipitation (MAP) Data for the Zambezi River Basin

Sub-basin	MAP (mm)
Kabompo	1,211
Upper Zambezi	1,225
Lungue-Bungo	1,103
Luanginga	958
Barotse	810
Cuando/Chobe	797
Kafue	1,042
Kariba	701
Luangwa	1,021
Mupata	813
Shire River/Lake Malawi	1,125
Tete	887
Zambezi Delta	1,060
Grand Mean	956

Source: World Bank (2010)

The rain season is unimodal and longer in the north and north-eastern compared to the south and south-western parts of the Basin. In addition to the ITCZ, additional synoptic systems that bring rainfall to the Zambezi Basin are the Zaire Air Boundary (ZAB) or Congo Air Mass and Tropical Cyclones as they veer away from their east to west track in the Mozambique Channel. A detailed analysis of historical and projected climate (temperature and precipitation) is presented in Chapter 5 of this volume. The mean annual precipitation in the Zambezi River basin is 956 mm year^{-1} or about 1,300 km^3 of water. It is estimated that less than 10 per cent of the precipitation flows through the river and reaches the Indian Ocean, the rest evaporates and returns directly to the earth's atmosphere (Euroconsult Mott MacDonald 2008). Zambezi tributaries peak rapidly after rainfall events and they reach their maximum discharge between January and March and decrease to minimum flows in October to November during the dry season. However, the Zambezi headwaters, the Kafue and Shire Rivers have large flood plains that delay peak discharge until later in the wet season or early in the dry season (Beilfuss 2012). Potential evapotranspiration also varies significantly across the basin. Annual potential evapotranspiration values vary from 1,000 mm to almost 2,000 mm with an average ranging between 1,560 and 1,600 mm (Euroconsult Mott MacDonald 2007; Beilfuss 2012).

The mean annual runoff is about 110,732 km^3 (3511 m^{3}s^{-1}) at the outlet of the Zambezi (Beilfuss 2012). The major contributors to the runoff are the northern sub-basins of Upper Zambezi, Kafue, Luangwa and Shire/Lake Nyasa. The mean annual runoff represents the expected annually available surface water resources of a basin. The available surface water resources of a basin is the amount of water generated through precipitation, over the entire basin which is equivalent to the water flowing in a river at a given terminus point over a specified period (Euroconsult Mott MacDonald 2008). This flow at a point on the river represents the integrated runoff for the entire basin above the point and if that point is the outlet of the basin or sub-basin, then that amount represents the available water resources for the basin. It is normally measured in volume of water over the period or an equivalent depth of precipitation over the basin. Different authors have published varying results of mean annual runoff for the Zambezi Basin (Euroconsult Mott MacDonald 2007; World Bank 2010; Beilfuss 2012). The differences are attributed to the errors made in computing some of the sub-basin catchment areas (World Bank 2010; Beilfuss 2012). The mean annual runoff of 110,732 km^3, for the entire Zambezi River basin, when compared to the mean annual precipitation of 1,300 km^3 translates into a runoff coefficient (efficiency) of 8.5 per cent. The Zambezi River has three distinct reaches with different characteristics. The Upper Zambezi Region from its origin up to Victoria Falls, the Middle Zambezi from Victoria Falls to Cabora Bassa and the Lower Zambezi from Cabora Bassa to the delta. The Larger Upper Zambezi is still the most natural part of the river largely due to absence of large reservoirs (McCartney et al. 2013). Below Victoria Falls, the river is highly regulated by the Kariba and Cabora Bassa dams which were mainly constructed for hydropower generation.

Operations of the two dams have resulted in a modified flow regime, i.e. attenuated peak flows and increased dry season flow, causing changes in water quality, reduction in flood plain areas as well as changes in the morphology and ecology of the river. The regulation has also caused changes to the composition of fish species and biodiversity in general, with the rise of tilapia and Kapenta

(Tanganyika sardine) fisheries in Lake Kariba and the reservoir at Cabora Bassa, and economic damage to the freshwater and marine shrimp fisheries in the delta (Euroconsult Mott MacDonald 2008). For more detailed discussions on ecological trends within the basin, see Chapters 3 and 4.

Groundwater

Groundwater resources in the Zambezi River Basin, like in the rest of the Sub-Saharan Africa region, are mainly derived from four main aquifer systems, namely: the low yielding but extensive Pre-Cambrian Basement Complex aquifer, consolidated sedimentary rocks such as sandstones and limestones, unconsolidated sediments, and volcanic rocks (Macdonald and Davies 2000). The Basement Complex aquifer is the main source of rural water supply in the Zambezi River Basin, providing potable water to the rural poor through boreholes fitted with Afridev hand pumps (Plate 6.1). In cases where yields are high, the aquifer has been exploited to provide water to institutions such as health centres and schools (Plate 6.2). While the quality of groundwater resources is generally acceptable for domestic consumption, water quality problems are mainly associated with high salinities, especially where the hydraulic gradients are very low. Other groundwater quality problems associated with drinking water supply within the basin include the presence of iron, sulphate, and fluoride concentrations above acceptable limits.

In Sections 2-5, the Hydrology of the Zambezi River Basin is discussed in detail, with a focus on four regions of the basin, namely: the Upper Zambezi Region, the Manyame Catchment (which forms part of the Tete Sub-basin), the Shire Sub-basin, and the Songwe and Ruhuhu catchments which empty their waters in Lake Malawi/Nyasa/Niassa.

Plate 6.1: A borehole in Thyolo District, Malawi, fitted with an Afridev hand pump

Plate 6.2: A borehole at Neno Health Centre, Malawi, fitted with a motorized pump, supplying water to the facility through an overhead tank

Hydrology of the Upper Zambezi Region
Characteristics of the Upper Zambezi Region

The Upper Zambezi Region covers an area of about 515,008 km^3 and consists of the sub basins of Kabompo, Upper Zambezi, Luanginga, Barotse, Lungue-Bungo and Cuando-Chobe. The Upper Zambezi Sub-basin (Zambezian headwaters) is part of the ecoregion of Savannah-dry forest which also covers the headwaters of Okavango and Kafue Rivers. It is bio-regionally outstanding with conservation status of nationally important and relatively intact aquatic ecosystems (Timberlake 2000). The rivers in this ecoregion are typically tropical, perennial, steep in some areas and include extensive network of grassy dambos (i.e., seasonally water-logged areas).

The Savannah-dry forest eco-region also covers the headwaters of all the other sub-basins in the region except Barotse which falls under the Upper Zambezi flood plains eco-region. This ecoregion is also treated as nationally important and relatively intact aquatic ecosystem. It is characterized by shallow and alluvial basin whose gentle slope and moderate rainfall have supported development of swamps and flood plains. The Upper Zambezi flood plains ecoregion begins at the confluence of Lungue-Bungo, Zambezi and Kabompo rivers and extends up to Victoria Falls. The ecoregion includes the vast and renowned Barotse flood plain which is about 200 km long and 40 km wide. The Barotse plains delay the flooding water for about 4-6 weeks (Beilfuss 2012).

Rainfall and Runoff in the Upper Zambezi Region

The mean annual precipitation in the Region varies from 797 mm in the Cuando-Chobe to 1,225 mm in the Upper Zambezi (Table 6.1). The mean annual precipitation for the region is 1,000 mm. About 50 per cent of the annual rainfall infiltrates into shallow aquifers which contributes to the base flow (Sharma and Nyumbu 1985). Potential evapotranspiration ranges from 1,666 mm per year in Luanginga to 1,337 mm per year in Kabompo (Beilfuss 2012). Flows begin rising during the wet season from December to January and increases rapidly from February to April. The recession takes place during the subsequent dry season and reaches the minimum in November. On average, the region contributes 37.249 km³ (1,181 m³s⁻¹) of water annually into the Zambezi, which is equivalent to 28.55 per cent of the total Zambezi runoff (World Bank 2010). The runoff in the region varies considerably with the coefficient of variation of 0.40. However, during drought years the magnitude and duration of averaged peak flows may be reduced by 70 per cent or more (Beilfuss 2012). Figure 6.2 shows the mean monthly discharge from the Upper Zambezi Region (at Victoria Falls) during average and drought years.

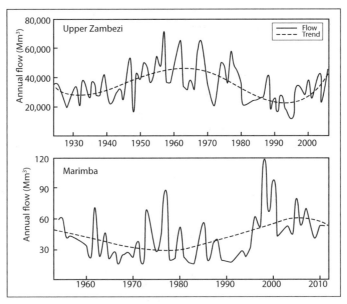

Figure 6.2: Mean monthly discharge from Upper Zambezi (at Victoria Falls) during drought and average years
Source: Beilfuss (2012)

Most of the flow in the Upper Zambezi Region is contributed by the Upper Zambezi Sub basin that generates 23.40 km³. The runoff efficiency value stands at 0.21, the highest in the region. The Kabompo, Lungue-Bungo and Luanginga sub-basins contribute 8.61 km³, 3.59 km³ and 2.19 km³ respectively. The Cuando-Chobe contributes nothing and Barotse generates a net loss of 0.56 km³ owing to the presence of a large wetland (Barotse Dampo flood plain) which attenuates peak flow leading to high evaporation rates (World Bank 2010). The Cuando-Chobe Sub-Basin also contains a wetland (Chobe wetland) which further attenuates the cumulative inflows from Barotse and its own headwaters but does not result in loss of cumulative inflows. Zambezi River Basin is characterized by extreme climatic variability that results in cycles of extreme hydrological events such as floods and droughts with devastating impacts to the people and the economy. Both seasonal and annual rainfall amounts are characterized by considerable variation such that floods and droughts of several years' duration have been recorded almost every decade (Beilfuss 2012). The biggest flood in Upper Zambezi happened in 1957/58 hydrological year when the annual runoff reached 72,800 Mm³ and the driest hydrological year was 1995/96 when the annual flow dropped to 12,300 km³. Figure 6.3 shows the annual flows hydrograph at Victoria Falls.

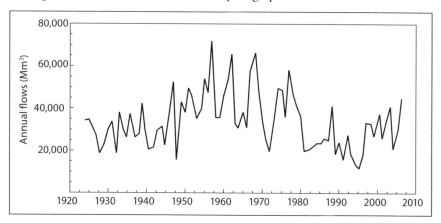

Figure 6.3: Hydrograph of annual flows for Zambezi River at Victoria Falls (Big Tree Station)

Hydrology of the Lake Malawi/Nyasa and Songwe and Ruhuhu Rivers

Characteristics of Lake Malawi/Nyasa and Ruhuhu/Songwe Catchments

Lake Malawi/Nyasa has a drainage basin area of about 98,700 km². The basin is shared among Tanzania (28 per cent), Malawi (64 per cent) and Mozambique (8 per cent). Despite the fact that the Tanzania part of the basin is less than one third (1/3) of the whole basin area, it contributes about 52 per cent of inflows into the lake. The basin is generally mountainous and hilly, with Mountains Rungwe and Kipengele being the highest peaks. The peak of Rungwe is 2,980 masl. Kipengere and Mporoto mountain ranges are actually the oldest plateaus found at the highest levels from 1,800 to 3,000 masl. The lowest lying area is around the lake which is about 470 masl. The elevated plateaus are characterized by rolling hills with steep-sided drainage channels cut into deeply weathered basement. Igneous intrusions form steep rocky hills which are prominent south of Njombe, just north of Songea and north of Mbamba Bay (SMEC 2015a).

Ruhuhu and Songwe Rivers drain two of the 10 main catchments that form the Lake Nyasa Basin on the Tanzania side. The Songwe River is an international boundary as it forms the border between Malawi and Tanzania while Ruhuhu catchment is the largest (Shela 2000a). The other eight catchments are Kiwira, Mbaka, Lufirio, Rumakali, Lumbira, Nkiwe, Mchuchuma and Mbawa. All the rivers draining the lake catchments are perennial. The catchment areas of the two rivers are 14,200 km² and 2,490 km² and contribute 37 per cent and 10 per cent of the inflows into the lake respectively. Lake Malawi/Nyasa is an ecoregion defined by its drainage basin. The ecoregion is classified as large lake ecoregion which is globally outstanding and vulnerable. The lake is the southern-most of the deep water lakes of the East African Rift Valley. The lake area is about 29,600 km² which makes it the eighth largest in the world and the only large natural lake in the Zambezi Basin (Timberlake 2000).

Rainfall and Runoff in Ruhuhu and Songwe Catchments

Lake Malawi/Nyasa Sub-basin is one of the areas with highest rainfall in the Zambezi Basin ranging from just below 1,000 mm year^{-1} at some points in Ruhuhu Catchment to more than 2,600 mm/year in the Songwe Catchment at Kyela. The average rainfall in the sub-basin is 1,400 mm year^{-1}. The rainy season starts in November, peaks in March and ends in late April or early May (Shela 2000b; SMEC 2015a). The river flows begin rising soon after the start of the rainy season with the recession observed from late April or early May. The potential evapotranspiration varies from about 1,478 mm year^{-1} in elevated areas to 1,687 mm/year in low-lying areas near the lake with the mean potential evaporation of 1,420 mm year^{-1} (SMEC 2015a). The mean annual runoff is 4.251 Mm³ and 1.365 Mm³ which translate into runoff coefficients of 0.27 and 0.29 for Ruhuhu and Songwe respectively. The runoff generation capacity of Ruhuhu and Songwe catchments is the highest in the Zambezi Basin reflecting the runoff generation capacity of Lake Malawi/Nyasa Basin as a whole. Table 6.3 shows the mean annual rainfall, runoff and evapotranspiration from some of the catchments of Lake Malawi/Nyasa Sub basin in Tanzania. It can be seen that Ruhuhu catchment is the largest contributor of inflow into the lake followed by Songwe.

Table 6.3: Mean annual rainfall, runoff and evapotranspiration for selected catchments of Lake Malawi/Nyasa Basin in Tanzania

River or Catchment	River Station	Area (km²)	Rainfall (mm year⁻¹)	Runoff (Mm³year⁻¹)	Evapotranspiration (mm year⁻¹)
Songwe	Kasumulu	3,550	1,310	1,365	925
Kiwira	Kiwira	187	2,064	174	1,132
Kiwira	Natural Bridge	709	2,134	712	1,130
Rumakali	Homage	414	1,977	351	1,129
Ruhuhu	Masigira	2,680	1,409	1,021	1,028
Ruhuhu	Kikonge	13,490	1,226	4,251	911

Source: SMEC (2015a)

Flooding has been an issue in various parts of Lake Malawi/Nyasa Basin particularly in the Songwe catchment due to high rainfall regime but also low conveyance capacity of the river channels and relatively flat slope of less than 1 per cent (SMEC 2015b). The high flooding areas are the lower parts of Songwe, Kiwira, Mbaka and Lufirio catchments. The Kyela flood plain normally inundates Kyela Town, and Songwe River meandering due to floods causes tension since the river forms the boundary between Malawi and Tanzania. Lahmeyer International (2013) observed that flooding normally occurs between March to May and stays for varying duration ranging from 10 hours, few days and weeks for water to disappear. Based on the analysis in the Lower Songwe, flooding which is defined for this river as peak discharge, greater than $250 \text{ m}^3 \text{ s}^{-1}$, does not occur every year. In the flow time series from 1986-2011 (25 years) there was a total of 9 years that flooding did not occur. On average, large floods that can cause some damage occur about once in 3 years.

Hydrology of the Shire River Sub-Basin

Characteristics of the Shire River Sub-Basin

The Shire River originates from the outlet of Lake Malawi/Nyasa at Samama in Mangochi area and runs through the Southern parts of Malawi for about 410 km to its confluence with Zambezi River at ZiuZiu in Mozambique (Shela 2000b; Government of the Republic of Malawi (GoM) 2013). The Shire is characterized into upper, middle and lower reaches mainly based on channel gradient. The Upper Shire has a channel bed drop of about 15 m over a 130 km distance from the outlet to Matope (Shela 2000b; Government of the Republic of Malawi [GoM] 2013). However, the uppermost reach from Mangochi to Liwonde is almost flat with a fall of 1.5 m over 87 km. The relatively flat feature of this reach makes it viable for flow regulation. From the outlet, the Shire River flows into Lake Malombe, a shallow floodplain of about 30 km and 15 km wide, located about 8 km south of Mangochi. Kamuzu Barrage is located at Liwonde on the downstream end of the Upper Shire. The barrage regulates flow of water in the river by storing water upstream and therefore helping to ensure year-round downstream flows to sustain key economic activities. The water level at Liwonde is only 2 m lower than the Lake Exit and operations of the barrage results in even lower gradient and reduced stream velocity. The low stream velocity coupled with flat landscape causes the river to meander creating a network of pools and channels whereby adjacent land is normally flooded during the rainy season (SMEC 2013b). The landscape of Upper Shire is typical of the Great African Rift Valley. The valley consists of gentle foot-slopes which rise to an altitude of approximately 700 m.

The Middle Shire begins downstream of Liwonde at Matope and drops 370 m in elevation over a distance of 80 km between Matope and Kapichira. The huge drop makes the reach viable for hydropower generation. The reach contains a series of gorges, falls and cataracts, namely Kholombidzo, Nkula, Tedzani, Mpatamanga, Hamilton and Kapichira. Nkula, Tedzani and Kapichira have been dammed for hydropower generation (Government of the Republic of Malawi (GoM) 2013). Other features of the river reach are meandering, rapids flanked by islets and abandoned river channels. Several tributaries join the main reach but the most important ones are the perennial Lisungwe and Mkurumadzi rivers. The Middle Shire valley forms the floor of the Great African Rift Valley and consists of undulating but rugged and densely dissected country. The Rift Valley is separated from the African surface by a scarp zone of broken terrain incorporating an altitude gain of several hundred metres (SMEC 2013b).

The Lower Shire Basin is a wide flat alluvial plain stretching a distance of 200 km from Kapichira Falls. The reach is sometimes further characterized into upper and lower sections. The upper Lower Shire runs for about 80 km from Kapichira (80 masl) to Chiromo where the elevation drops by 35 m (Shela 2000a). This section is dominated by the seasonally flooded Elephant Marsh. The Ruo River joins the Shire where the lower section begins up to the confluence with the Zambezi River (30 masl). The section is occupied by another extensive wetland known as the Ndindi marsh. The Lower Shire Valley is a continuation of the Rift Valley whose floor is flanked by escarpments associated with major fault lines which follow a south-east to north-west trend. Other major tributaries that join the main river in the lower section are Mwanza and Nkombedzi-wa-Fodya. The tributaries normally flood heavily during the rainy season.

Rainfall and Runoff in the Shire River Sub-Basin

The Upper Shire is relatively dry with mean annual rainfall of about 700 mm (SMEC 2013b). The rainy season runs from November to April when more than 90 per cent of the rain falls. The May to October period constitutes the dry season with less than 10 per cent of the rain. The river flow is heavily dominated by outflow from Lake Malawi/Nyasa. The outflow from the lake begins at about 471.5 masl and it accounts for almost all the flow in the dry season (SMEC 2013b). Long-term variations in the lake levels (Figure 6.4) have caused periodic changes in the flow characteristics of the river. The river ceased to flow completely from 1915 to 1934 when the lake level reached 469 masl. The low lake levels during the period in question are attributed to low rainfall in the catchment prior the 1915 period. From 1935 to 1937, the increasing lake levels finally pushed the water through the sand bars, debris and reopened the river channel at its mouth and along its upper reaches that were blocked by vegetation growth and pilling of sedimentation from small tributaries near the lake. The mean annual runoff for the Upper Shire measured at Matope is 405 m^3s^{-1} where the catchment area corresponds to 7,200 km^2.

The frequent rise and fall of lake levels indicate that it is very sensitive to climatic variability associated with higher rainfall and drought periods respectively. The computation of the water balance of Lake Malawi/Nyasa is complicated by lack of adequate stream flow and rainfall data. Most of the streams are not gauged and the gauged ones have short records with a lot of gaps. Rainfall is only measured on land areas surrounding the lake and lake level/outflows are influenced by regulation of the Shire River (Shela 2000b).

The Middle Shire has slightly higher mean annual rainfall than the upper part with 700-800 mm. The river flow in the reach is heavily regulated by the Kamuzu Barrage at Liwonde that was constructed in 1965. The barrage also provides limited control of water levels upstream in the Shire River and Lakes Malombe and Malawi/Nyasa. It is mainly operated to satisfy the requirements of electricity generation at Nkula Hydropower Station (124 MW) and other downstream cascade of hydropower stations. Currently, the total hydropower generation in the Shire River is 200 MW whereas the potential is 400-500 MW. Before the construction of the Kamuzu Barrage, a natural sandbar at Shire inlet from Lake Malawi/Nyasa has historically controlled outflow from the lake (Beilfuss 2012).

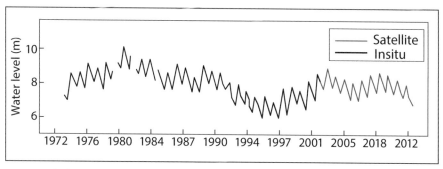

Figure 6.4: Water levels of Lake Malawi/Nyasa 1972-2012. Modified from Faraji (2016)

The Kamuzu Barrage operates at lake levels of 473.2 and 475.32 masl and provides a firm flow of 170 m³s⁻¹. The operation of the barrage also benefits irrigation in the Lower Shire Valley. It has also indirectly facilitated maintenance of relatively high lake levels that have supported navigation and fisheries sectors (Shela 2000b). Despite the regulation, the flow regime of the Middle Shire exhibits greater variation between peak and low flows than the upper reach. In normal years of lake outflows, the rainy season flow is about one and half to three times the dry season flow. Tributary inflows in this reach are also significant as they are estimated to contribute about 30 per cent of the shire water in the rainy season and less than 10 per cent in the dry season (SMEC 2013b).

Rainfall in Lower Shire ranges from 700-1,000 mm which is far below the flood plain evaporation estimated to be 2,000 mm per year (Beilfuss 2012; SMEC 2013b). Flow in the whole reach is generally slow due to the flat flood plain. The Ruo River running from Mozambique joins the Shire downstream of Chiromo. The river is the largest of the Shire tributaries and in some years it actually has a greater impact on the water table in the marsh than the Shire. The Ruo River can carry as much as 3000 m³s⁻¹ into the Shire during peak floods. The mean annual flow of the Shire River at Chiromo is 480 m³s⁻¹ where the catchment area is 18,240 km² (SMEC 2013b).

Overall, the Shire River/Lake Malawi-Nyasa Sub-basin is the second largest contributor of runoff into the Zambezi after Upper Zambezi. The mean annual precipitation and mean annual potential evapotranspiration for the entire Shire catchment is 965 mm year⁻¹ and 1,608 mm year⁻¹ respectively (DHI 2015).

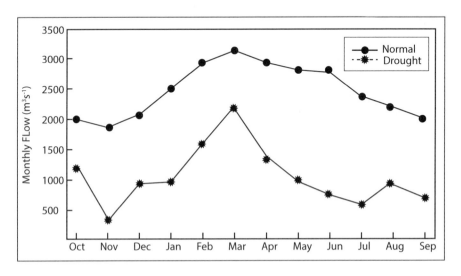

Figure 6.5: Mean monthly discharge from Shire River/Lake Malawi–Nyasa Sub–basin during drought and average years. Modified from Beilfuss (2012)

The catchment is a typical representative of the Zambezi Basin in terms of mean rainfall and evapotranspiration save for the runoff which is influenced by outflows from the lake. Mean annual runoff of the sub-basin is estimated to be 15,700 Mm^3 (a flow rate of 498 m^{3}s^{-1}) which is about 20 per cent of the Zambezi Basin total (World Bank 2010). Despite its high contribution to the total runoff and being a headwater catchment, the sub-basin has a low runoff efficiency of 0.09 only. This is due to large attenuating effects of the lake and the downstream floodplains. Peak runoff is attained between February and April. When levels of the lake are highest and tributary inflows into the shire are substantial, large floods can occur in the Lower Shire. Flooding events of up to 18,150 Mm^3 have been observed. In drought and/or low lake levels years, the flow falls to about 9200 Mm^3 (Beilfuss 2012). Figure 6.5 shows the mean monthly flows in the Shire River/Lake Malawi-Nyasa Sub-basin during average and drought years.

Hydrology of the Manyame River Catchment
Characteristics of the Manyame River Catchment

Manyame is part of the Tete Sub-basin and has a catchment area of 40,497 km^2. The Manyame River qualifies as an international river as it transverses the Zimbabwe and Mozambican border. It is called Manyame in Zimbabwe and Panhame in Mozambique. The Manyame River is also important as it is a significant source of runoff along the Zambezi reaching Cabora Bassa reservoir. The Manyame River Catchment originates near Marondera area that is located 80 km to the east of Harare. The river stretches for more than 400 km before it discharges into the Zambezi River upstream of the Cabora Bassa reservoir (Makhanga 2011; Motsi et al. 2002). The upper and lower parts of the catchment have distinct characteristics which normally necessitate description of the catchment based on the two subdivisions and sometimes the middle part is also considered. The Upper Manyame catchment lies at an altitude ranging from 1,400 to 1,500 masl and the lowest altitude is about 1,000 masl. The catchment is defined as the portion from the source to the Manyame Dam which has a catchment area of 3,930 km^2. The catchment area is a gently undulating, featureless plateau with a few hills, particularly around Lake Chivero (Motsi et al. 2002). The major tributaries (of the upper Manyame Catchment) are Nyatsime, Ruwa, Mukuvisi and Marimba rivers. Manyame Dam (490 Mm^3) is the biggest dam in the catchment but three other relatively big dams are located in the upper catchment along the main stem, namely Lake Chivero (250 Mm^3), Seke (3.65 Mm^3) and Harava (9.25 Mm^3). The reservoirs created by these dams are largely used for irrigation and water supplies of the urban centres of Harare, Chitungwiza, Ruwa and Norton. They are operated to maximize storage of water to meet the demands. Downstream of Lake Manyame, two other relatively large dams exist, namely Biri (172 Mm^3) on the main stem and Mazvikadei (360 Mm^3) located on Mukwadzi River.

Upper Manyame is the most urbanized catchment in Zimbabwe (Mwedzi et al. 2016a) and certainly one of the highly urbanized areas in the Zambezi Basin. The urban centres are major sources of pollution due to discharge of domestic and industrial effluents. The effluent is discharged in both forms, treated and untreated (Nhapi 2009). Agricultural activities and rural settlements are also present in the catchment. Water is abstracted from rivers and other small impoundments for irrigation and livestock rearing purposes. However, poor management of farms and agricultural inputs by some farmers has resulted in siltation of water bodies and contamination by nutrients and chemicals. It is due to pollution from urban centres and agricultural activities that give Upper Manyame the reputation of probably being the most polluted in Zimbabwe (News Day 2011). The Lower Manyame is predominantly a rural catchment where major water demands are mainly for agricultural purposes. The catchment is relatively undeveloped whereby the amount of storage available is about half of that of Upper Manyame.

Stream-flow characteristics at different sections in the Manyame catchment (upper and lower) are largely dependent on rainfall and the presence, position and distance from dams. The Manyame Catchment is therefore hydrologically fragmented with the stream-flow characteristics upstream of dams being different to that immediately downstream of dams. The extent of the fragmentation in the catchment, however, is minor as the greater part of the rivers remain free-flowing and the downstream impacts are only limited to the first 20 km after which the downstream effects of the dams are diminished. This is mainly because dams are limited to the upper Manyame Catchment which receives less rainfall and is smaller than the Lower Manyame. The main changes entail a continued replacement of high flows, floods and minimum flows by extreme low flows and an increased number of zero flow days downstream of dams. Upstream changes entail a continued alteration and increase in high flows (Mwedzi et al. 2016b).

Rainfall and Run-off in the Manyame Catchment

Rainfall in the Manyame Catchment is seasonal falling between November and March, with peak rains coming in January. Mean annual precipitation in the Manyame Catchment is 768 mm (see Table 5.2 in chapter 5). The mean annual precipitation for Harare is 817 mm (Motsi et al. 2002). The river flow is seasonal where most of the flow is confined to the rainy season (mid-November-March). The Mean annual runoff of the Manyame Catchment decreases as you move from Upper Manyame Catchment (where the value ranges from 101-150 mm year^{-1}) to the Lower Manyame Catchment (where the value ranges from 51-100 mm year^{-1}). The coefficients of variation of the annual flows increase from Upper Manyame (75-100 per cent) to Lower Manyame (101-125 per cent) (Mazvimavi et al. 2007). The runoff efficiency can be estimated to vary from (0.126-0.183) for

Upper Manyame to (0.065-0.122) for Lower Manyame. The mean annual runoff for the entire catchments is about 0.834 km³ (834 mm³) which approximates to 0.75 per cent of the Zambezi (World Bank 2010).

Trend Analysis of Discharge in the Zambezi River Basin
Discharge Data Used in Trend Analysis

Discharge data from a total of 13 river gauging stations were obtained from different sources (Table 6.4). Two stations were obtained from the Water Resources Department of the Ministry of Water Development and Irrigation of Malawi; three stations were obtained from the Lake Nyasa Basin Water Board of the Ministry of Water and Irrigation of Tanzania; one station was obtained from the World Wide Fund for Nature of Zambia; and seven stations were obtained from the Zimbabwe National Water Authority. Trend analysis requires that the record length of discharge data be of at least 20 years for meaningful results and discussion. The analysis further requires that there should be no data gaps equivalent to 10 per cent or more in the record of time series being considered (Valimba 2004). Preliminary analysis of data indicated that one station did not qualify for further analysis owing to its short record period. The station is Mukwadzi at Mazvikadei Dam. Table 6.4 summarizes the details of the river discharge stations used for trend analysis.

Table 6.4: Summary of river gauging stations data used for trend analysis

Catchment	Station Name	Latitude	Longitude	Record Length
Manyame	Mukwadzi	17°05' S	30°18' E	1971-2011
Manyame	Chinhoyi old road bridge	17°21' S	30°13' E	1965-2011
Manyame	Seke Dam D/S	17°59' S	31°04' E	1951-2011
Manyame	Seke Dam U/S	18°00' S	31°07' E	1958-2011
Marimba	Chivero Dam U/S	17°55' S	30°52' E	1953-2011
Manyame	Chivero Dam	17°53' S	30°46' E	1953-2011
Mukwadzi	Mazvikadei Dam	17°23' S	30°29' E	1989-2010
Upper Zambezi	Victoria Falls-Big tree	-		1924-2006
Shire	Mangochi	-		1975-2004
Shire	Liwonde	-		1948-2006
Ruhuhu	Masigira	-		1971-2015
Songwe	Kasumulu	09°56' S	35°11' E	1965-2011
Songwe	Itumba	-		1958-2011

Results of Trend Analysis

The statistic (Mann-Kendall) used to calculate the trend analysis is described in Chapter 2 of this volume. The trend analysis was initially conducted for the 1924-2006 period for which Victoria Falls station in Kariba Sub-basin, Zambia, has records. The result indicated an insignificant decreasing trend at 99 per cent, 95 per cent and 99 per cent confidence levels. The analysis was repeated for the 1958-2006 period in order to show the effect of using piecemeal record in trend analysis whereby Victoria Falls demonstrated a significant decreasing trend at all 3 confidence levels, Marimba at Chivero Dam station, a tributary of Manyame Catchment, demonstrated a significant increase at all levels and Shire at Liwonde station demonstrated an insignificant decrease at the 3 levels. For the 1965-2006 time period, Victoria Falls demonstrated a significant decrease for the 99 per cent and 95 per cent confidence levels only, Marimba demonstrated a significant increase at all levels and Songwe at Kasumulu station demonstrated an insignificant decreasing trend at all levels. The final analysis was conducted for the 1971-2006 period where Victoria Falls demonstrated an insignificant decreasing trend at all levels. Marimba demonstrated a significant increase at 95 per cent and 90 per cent levels; Shire demonstrated a significant decrease at 90 per cent and 95 per cent levels while Songwe and Manyame demonstrated an insignificant decreasing trend at all levels. Table 6.5 summarizes the trend analysis results. Other stations could not be analyzed due to insufficient data that is indicated by the computer programme results as (-9.9).

Analysis of trends using piecemeal records may give different results for the same station as indicated in the foregoing discussion. Normally, the start and end year for the analysis needs to be the same to allow a homogeneous comparison of results (Valimba 2004). However, for this case the purpose was to calculate the long-term flow trend and therefore the need to utilize the available record as much as possible. The choice of the start year of analysis was made such that it coincides with the start of at least 2 stations while the end year was fixed up to 2006 for which at least most of the stations in the five catchments under consideration had records.

The long-term trend of river flows demonstrated by 3 river catchments of Songwe, Shire and Upper Manyame and the sub-basin of Upper Zambezi is that there was an insignificant decrease for the time period up to 2006 save for the significant increasing trend for Marimba (Upper Manyame tributary) which is attributed to changes in land cover due to urbanization in the catchment (Gumindoga et al. 2014). The decreasing long-term trend was highest in Songwe ($Z = 1.59$) followed by Upper Manyame ($Z = 1.58$), Shire ($Z = 0.72$) and Upper Zambezi Region ($Z = 0.71$). The rainfall trends discussed in Chapter 5 (this volume) supports the river flows trend findings. Rainfall in Songwe, Shire and Manyame was shown to have an insignificant decreasing trend. Although the

Barotse catchment which is part of the Upper Zambezi region has an insignificant increasing rainfall trend, the contribution of this to flows would not affect the larger sub-basin due to the fact that hydrologically, the Barotse catchment generates a net loss due to high evaporative losses. The Songwe, Ruhuhu, Upper Manyame and Upper Zambezi catchments have similar hydrological characteristics of being headwater catchments with relatively higher runoff generation capacities than the Zambezi Basin average. Despite the similarity, the catchments have strikingly different developments. Upper Manyame is a highly urbanized area with commercial farming practices (Motsi et al. 2002; Mwedzi et al. 2016a), Songwe and Ruhuhu are rural catchments with unsustainable land management practices leading to high degradation and sedimentation that threaten even the stability of the river itself (Lahmeyer International 2013; SMEC 2015b), and Upper Zambezi is a relatively undisturbed region in the Zambezi Basin (McCartney et al. 2013).

Table 6.5: Results of trend analysis for some of catchments/sub-catchments of Zambezi Basin

Catchment	Zambezi	Ruhuhu	Songwe	Shire	Manyame	Manyame	Manyame	Marimba
Station	Victoria Fall	Masigiri	Kasu-mulu	Liwonde	Mukwadzi	Seke Dam	Chivero	Chivero
1924-2006	-0.7	-9.9	-9.9	-9.9	-9.9	-9.9	-9.9	-9.9
1958-2006	-2.89	-9.9	-9.9	-0.72	-9.9	-9.9	-9.9	2.26
1965-2006	-2.68	-9.9	-1.59	-1.27	-9.9	-9.9	-9.9	3.04
1971-2006	-0.72	-9.9	-0.01	-2.26	-9.9	-1.58	-9.9	2.35
Z (99 %)	2.57	2.57	2.57	2.57	2.57	2.57	2.57	2.57
Z (95 %)	1.96	1.96	1.96	1.96	1.96	1.96	1.96	1.96
Z (90 %)	1.67	1.67	1.67	1.67	1.67	1.67	1.67	1.67

*(-9.9 means that the programme could not calculate trend results due to insufficient data)

Trend analysis also reveals that long-term flows follow cyclic or oscillating patterns (Figure 6.6). The Upper Zambezi region at Victoria Fall depicts an oscillation frequency of about 60 years. Marimba River does not have long enough record to display the full oscillation but the half cycle indicates duration of about 28 years. During periods or cycles of low rainfall the flows are certainly low but the amount of rainfall that appear as runoff decreases due to lowering of groundwater tables which leads to decrease in runoff generation capacity of the catchments (Mukosa et al. 1995). Likewise, during cycles of high rainfall the high flows are even higher due to prevalence of higher groundwater levels that contribute as base flows (Mazvimavi and Wolski 2006) which leads to the increase in runoff generation capacity of the catchments.

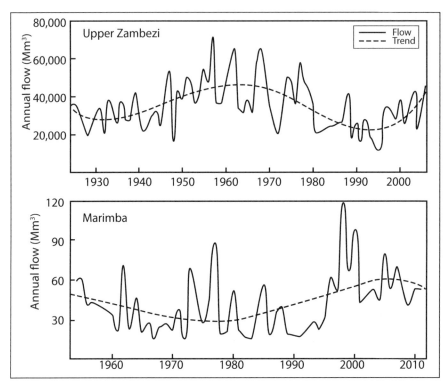

Figure 6.6: Trend analysis showing oscillating flow pattern for Upper Zambezi sub basin and Marimba Catchment

Discussion

The Zambezi River Basin climate has been described as one of the most variable of any major river basin in the world, with an extreme range of conditions across the catchment and through time (Beilfuss 2012). Rainfall varies from more than 1,600mm per year in some far northern highland areas to less than 550 mm per year in the low lying water-stressed southwestern portion of the basin (Beilfuss 2012). This variation in annual rainfall gives rise to unique patterns of runoff in each of the Zambezi sub-basins (Beilfuss and dos Santos 2001; Beilfuss 2012). The Basin Runoff follows the rainfall pattern and is therefore characterized by high seasonal and annual variability across the basin (Beilfuss 2012; McCartney et al. 2013). Runoff is thus concentrated in the northern part of the basin, where five major catchments contribute almost two-thirds of the total runoff. The entire basin (particularly the drier sub-basins) is therefore highly susceptible to droughts (often multi-year droughts) that occur in a cyclic manner-nearly every decade. These droughts are likely to become worse with climate change (Beilfuss 2012).

The observed average runoff efficiency of only 8.5 per cent across the entire basin means that, on average, only 85 mm runoff is generated annually from nearly 1,000 mm annual rainfall. This presents a major challenge in the hydrology of the catchment and makes it very vulnerable. Beilfuss (2012) further noted that small changes in rainfall produce large changes in the Zambezi runoff. The low runoff efficiency and high dryness index of the basin indicate high sensitivity to climate change (IPCC 1998). Although the severity of the impacts of climate change depended primarily on the magnitude of change, the hydrological sensitivity of the basin is also important (IPCC 1998). Climate change is expected to increase runoff variability, and consequently the vulnerability of the basin. In general, with climate change, the Zambezi is expected to experience drier and more prolonged drought periods, and more extreme floods. Multiple studies estimate that Zambezi runoff will decrease by 26-40 per cent by 2050, thereby increasing the water stress in the Basin (Beilfuss 2012).

Prior to damming, each Zambezi sub-basin contributed independently to the characteristic pattern of runoff in the catchment (Beilfuss 2012; Beilfuss and dos Santos 2001). However, these patterns have been altered by the construction of large dams for hydropower purposes, particularly Kariba and Cabora Bassa dams on the main stem, as well as Itezhi-Tezhi and Kafue Gorge Upper dams on the Kafue River tributary. The Upper Zambezi remains the most natural portion of the Basin as it does not have major dams on it (Beilfuss 2012; Beilfuss and dos Santos 2001). Runoff in the lower Zambezi basin is now mainly defined by regulated outflows from Cabora Bassa, partially regulated Shire River inflows and flashy runoff from the Mozambique plateau (Beilfuss and dos Santos 2001). These new patterns of runoff pose severe constraints on the potential for improving the delta's hydrological regime in the future as they have a profound effect on the flooding regime of the Zambezi Delta (Beilfuss and dos Santos 2001). The Zambezi Delta has thus undergone profound hydrological changes being transformed from a dynamic flood pulse system-maintained by runoff from a catchment extending over eight countries-to an isolated system dependent on local rainfall and runoff (Beilfuss and dos Santos 2001).

The operation of dams on the Zambezi River has resulted in an increase in dry-season flows and a delay and decrease in peak flows during the flood season (McCartney et al. 2013). Overbank flood pulses are reported to now occur only during major floods and are of inadequate volume and duration to sustain healthy flood plain systems of global importance, such as Kafue Flats, Mana Pools, and the Zambezi Delta (Beilfuss 2012). Furthermore, these flood pulses are often mistimed as they are generated during emergency flood releases or the late dry season in response to required drawdown releases. Constant dry-season flows generated from hydropower turbine outflows have replaced dry season flood-recession which was essential for river-dependent agriculture, fisheries, and wildlife (Beilfuss 2012). Lake Kariba is now important in curtailing floods and droughts which emanate from the upper Zambezi. Floods that occur upstream are absorbed and may not occur downstream. Consequently, hydrological extremes at Cabora Bassa are now greatly influenced by runoff accumulated within the middle Zambezi. Inflows to Cabora Bassa, therefore now resemble the characteristic pattern of natural inflows due to substantial unregulated runoff contributed from the Luangwa River catchment. These observed changes in magnitude, timing, duration, and frequency of flooding have severe consequences for the social and ecological health of the entire Zambezi Basin (Beilfuss and dos Santos 2001). Another problem presented by the Zambezi dams is that of water lost through evaporation. More than 11 per cent of the mean annual flow of the Zambezi evaporates from the large hydropower dams (Beilfuss 2012). The consequent average annual potential evaporation of about 1,560 mm far exceeds rainfall across the basin (Beilfuss 2012). For instance, water loss from the surface of Lake Kariba exceeds 2,000 mm per year (Beilfuss 2012). These water losses increase the risk of shortfalls in power generation, and significantly impact downstream ecosystem functions.

The trend analysis of flows at Victoria Falls station indicate that the time series had a shift in 1948 and 1980 (Laraque et al. 2001) as illustrated in Figure 6.7. The mean annual flows for the 3 resulting period segments, i.e.1924-1948, 1949-1980 and 1981-2006 are 30,189.6 Mm^3, 44,155.6 Mm^3 and 26,495 Mm^3 respectively according to the Bayesian method of Lee and Heghinian (1977), Buishand U test (Buishand 1984) and non-parametric test of Pettit (1979). The significant decreasing trends for the 1958-2006 and 1965-2006 time segments is explained by the fact that the 1949-1980 was a wetter segment than the earlier and latter segments (Figure 6.7), hence affecting the trend results. These results are collaborated by Beilfuss (2012) who analyzed the Victoria Falls station with a time series record from 1907-2006. He suggested that time series of annual flows reveal long-term cycles of high, medium and low runoff. The time series changed in 1945, 1981 and 1999 such that mean annual runoff for the 1907-1946 and 1982-1999 segments were lower than long-term mean flow. However, river flows upstream of Lake Chivero (on Marimba River) demonstrates a significant increase in runoff over the period 1956-2006. This increase is more pronounced from the 1990's going forward as shown in Figure 6.8. This agrees with findings by Gumindoga et al. (2014) who observed a 73.6 per cent increase in stream flows of the same river between 1980 and 2010. These increases coincided with decreases in woodlands and increases in urban areas. These findings are very relevant in the Zambezi River Basin countries, which are experiencing rapid urbanization but often lack planning and design (Gumindoga et al. 2014). Land-use change from natural catchments to peri-urban or urban areas has been reported to influence many processes of the hydrological cycle, such as interception, infiltration, evaporation and runoff processes (Dams et al. 2013; Furusho et al. 2013). The study therefore indicates that the Zambezi Basin hydrology is also under pressure from urbanisation and changing land uses.

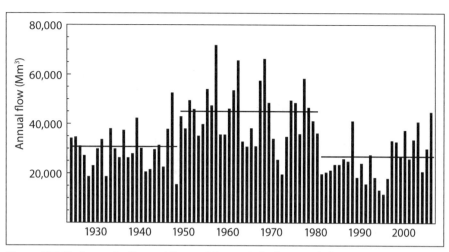

Figure 6.7: Hydrograph of annual flows at Victoria Falls showing shifting mean annual runoff

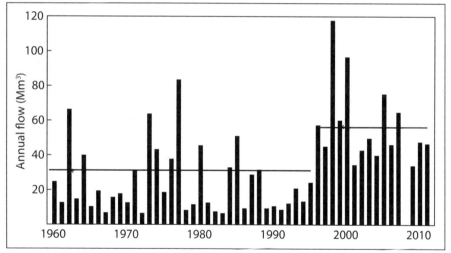

Figure 6.8: The hydrograph of annual flows for Marimba River in Manyame Catchment

Hydrological changes represented by decreasing flow trends are largely a result of consumptive water uses in the catchment, sub-basin and the basin at large. It was estimated by Euroconsult Mott MacDonald (2008) that around 20 per cent of the total Zambezi Basin runoff is used out of which energy sector constitutes the largest user whereby 16.46 per cent is lost through direct evaporation from reservoirs created for hydropower generation. Other sectors are irrigated agriculture (1.43 per cent), environmental and flood flow releases (1.16 per cent), rural domestic (0.02 per cent), urban domestic (0.17 per cent), industrial (0.02 per cent), mining (0.12 per cent) and livestock (0.11 per cent). Projected water uses due to population growth and increase of socio economic activities are expected to reach 37.32 per cent of the mean runoff by 2025 under high development scenario in Zambezi Basin. Given the projected runoff decrease of 26-40 per cent by 2050 due to climate change and even if an assumption is made that the projected water uses will be attained by 2050 instead of 2025, it means that water uses from the basin by mid-century will account for 63-77 per cent of the current mean annual runoff.

This level of water use in a basin is unsustainable and will no doubt exacerbate the poor condition of livelihoods dependent on goods and services derived from the flow-dependent ecosystem structure and functioning since its state will be heavily impacted. To make matters worse, most of the dams in the Zambezi Basin were designed without consideration for environmental flow releases and even where dams have been designed and required to provide ecological releases like the Itezhi-Tezhi Dam's prescribed freshet of 315 m^3s^{-1} for Kafue Flats, it has never been properly implemented (Beilfuss 2012). The situation calls for proper and just implementation of policies, legislations and regulations that promote social equity, environmental integrity and economic efficiency in utilization of water resources that is Integrated Water Resources Management (IWRM). Kaaya and Lugomela (2015) observed that problems facing deteriorating aquatic ecosystems were related to water resources management problems and that the need for protection of the resources was a key driver for adoption of IWRM approaches in Tanzania. Therefore, if proper policies and legal environment as well as institutional framework are put in place to operationalise IWRM, it can be a better prescription for water management problems that are facing the larger Zambezi River Basin.

The Zambezi River Basin IWRM Strategy (Euroconsult Mott MacDonald 2008) identified several issues and challenges that face the Basin with respect to water resources management. The identified issues are extreme variability and uneven distribution, the way resources are presently managed, predominance of internationally shared watercourses even for small sub-basins like the Songwe catchment, the importance of the aquatic environment (wetlands) and the extremely valuable ecosystem of the Zambezi Basin; and widespread poverty and low satisfaction of basic human needs like water, sanitation, energy and food. These issues/challenges were grouped into 4 water resources management (WRM) broad areas. Box 6.1 summarizes the four broad areas of key issues in WRM according to (Euroconsult Mott MacDonald 2008).

Box 6.1: Key Water Resources Management (WRM) issues in the Zambezi River Basin

Integrated and Coordinated Water Resources Development and Management| Environmental Management and Sustainable Development | Adaptation to Climate Variability and Climate Change|Basin-wide Cooperation and Integration.

Source: Euroconsult Mott MacDonald (2008)

Of the four issues, the Basin-wide Cooperation and Integration was crucial to start with since the prevalence of internationally shared watercourses in the Zambezi Basin was considered as one of the challenges in the management of water resources. Since as early as 1990s, independent approaches by riparian countries to conduct monitoring, assessment, planning, development, conservation and protection of water resources in their own territories, save for a few isolated cases, were seen as unsustainable (Shela 2000a; SADC 2005). Shela (2000a) argued for the Zambezi River Basin riparian states to seize the opportunity for joint water resources management. The SADC Regional Water Policy was promulgated in 2005 as one of the tangible steps towards regional integration in water resources management in the SADC block where Zambezi River Basin is the largest. In the policy, the Dublin Principles of IWRM were adopted as representing the best water resources management practice (SADC 2005). One of the key purposes of the regional water policy was to give guidance for harmonizing national water policies and management of water resources in member states. Likewise, the agreement on the establishment of the Zambezi Watercourse Commission was signed in 2004 as a result of long negotiations based on the convention of the Law of Non-navigational Uses of International Watercourses as well as the Revised Protocol on Shared Watercourses in the SADC adopted in August 2000. The key objective of the establishment of the commission is to promote the equitable utilization of the water resources of the Zambezi Watercourse as well as the efficient management and sustainable development thereof (SADC 2004). However, it is worthwhile to note that the agreement for the establishment of the Zambezi Commission came into force in June 2011 after ratification by member states and therefore the interim commission that was established after signing of the agreement changed into a permanent commission and moved its headquarters from Gaborone to Harare.

In response to the issues that were identified, Euroconsult Mott MacDonald (2008) developed four strategic objectives in order to address each of the issues with an overall objective of ensuring, *"equitable sustainable utilization of water for social and environmental justice, regional integration and economic benefit for present and future generations"*. Despite the Zambezi River Basin Strategy coming in 2008, and identifying Basin-wide Cooperation and Integration as one of the issues to be addressed, some regional cooperation efforts were happening for a little longer than 2 decades before that. The Zambezi River Action Plan (ZACPLAN) was put in place in 1987 by the Southern African Development Co-ordination Conference (SADCC), forerunner of SADC, with an objective of achieving environmentally sound planning and management of water and related resources in the Basin. The ZACPLAN was followed by the Zambezi Action Plan Projects 6 (ZACPRO 6) titled, Development of an Integrated Water Resources Management Plan for the Zambezi River Basin, which was implemented in two phases. Phase I ran from 1995 to 1999 with an objective of developing a knowledge base of water and related information to provide a sound basis for the planning and development of water resources of the Basin. Phase II started in 2001, as a follow-up to Phase I, with an objective of establishing an enabling institutional environment and management tools for IWRM. It was reformulated in 2003 to refocus on IWRM and facilitate social and economic development and protection against floods, droughts, water resources pollution and environmental degradation in the Basin (Euroconsult Mott MacDonald 2007; Euroconsult Mott MacDonald 2008). ZACPRO 6 Phase II culminated into the signing of the Zambezi Watercourse Commission agreement and the Zambezi IWRM Strategy. In these processes, the Zambezi River Authority (ZRA), a joint body for the management of the Zambezi between Zimbabwe and Zambia, played a key facilitating role. Other previous regional cooperation efforts worth mentioning is the 1995 Protocol on Shared Watercourses in the SADC, the forerunner of the 2000 revised protocol. Box 6.2 gives the strategic objectives identified by Euroconsult Mott MacDonald (2008).

It is obvious that most of the past efforts on regional integration and cooperation and the strategies/main actions for achieving the strategic objectives proposed in the Zambezi IWRM Strategy have produced tangible results on the fourth objective. The strategic objective on operationalization of institution frameworks in support of basin-wide water resources development and management has taken a significant leap with the ratification of the agreement on the establishment of the Zambezi Watercourse Commission and, consequently, the establishment of the permanent Zambezi Commission, normally abbreviated as ZAMCOM. However, past efforts on regional IWRM and Development have not had similar achievements. It is hoped that the establishment of a permanent ZAMCOM will advance the agenda on the same, though some riparian states have notable advances in realizing the implementation of IWRM approaches on their own. Tanzania, for example, promulgated a National Water Policy (NAWAPO) in 2002 which enshrined the IWRM principles even before the SADC Regional Water Policy came into being (United Republic of Tanzania [URT] 2002). The NAWAPO of 2002 was followed by a comprehensive Water Resources Management Act No. 11 of 2009 that, among other things, required all Basin Water Boards to prepare Integrated Water Resources Management and Development (IWRMD) Plans that will become blueprints for water resources management in the future (United Republic of Tanzania 2009). As a result, the Songwe and Ruhuhu catchments have Catchment IWRMD Plans that will be implemented as part of the larger Lake Nyasa Basin IWRMD Plan (SMEC 2015c).

Box 6.2: Strategic Objectives for addressing each of identified WRM issues in the Zambezi Basin

> ▶ Develop and manage water resources so as to serve social and economic development in the Basin.
> ▶ Mainstream environment in the development and management of water resources in the Basin.
> ▶ Adapt water resources management to current and future climate variability and change.
> ▶ Operationalize the institutional frameworks in support of basin-wide water resources development and management.

Source: Euroconsult Mott MacDonald (2008)

It is anticipated that proper implementation of the plans will lead to holistic and sustainable water resources management and development. The National Water Policy (2004) of Malawi has provisions related to water resources management. The policy requires that water should be managed and used efficiently and effectively in order to promote its conservation and future availability in sufficient quantity and acceptable quality. It further required that all programmes related to water should be implemented in a manner that mitigates environmental degradation and, at the same time, promote the enjoyment of the asset by all (SMEC 2013a). The new Water Resources Act of 2013 provides for the establishment of the National Water Resources Authority (NWRA) which would oversee national-level coordination of activities related to water resources management and development; and Catchment Management Committees (CMCs) which would provide for decentralized management, control and oversee the development of water resources at the catchment level. However, only a handful of CMCs are in place and the NWRA is not yet functional (DHI 2015). As such, Malawi does not have an institutional mechanism at the moment to allow the implementation of IWRM as well as coordinate integrated investment planning and system operation for the Shire River Basin (DHI 2015). This implies that the SADC Regional Water Policy's aim of harmonizing national water policies and managing water resources is yet to bear fruits. It is also notable that, although the ZAMCOM agreement has entered into force, Malawi did not ratify it.

In Zimbabwe, the Water Act of 1998 was a major piece of legislation that reformed water management, development and utilization of water resources in the country. The act, among other things, clearly stated that water management should involve all stakeholders to the lowest possible level and that the environment is considered to be a consumer which means that water quality control and environment protection were given due attention (Latham 2001). The act established the Zimbabwe National Water Authority (ZINWA) with the purpose of overseeing the planning, development and management of water resources as well as providing potable water to local authorities and government institutions. The act also established the Catchment Councils (CC) and Sub-Catchment Councils (SCC) which allowed participation of stakeholders at all levels, particularly lowest levels, in issues related to licensing of water abstractions (water permitting) and other water resources management and development issues. Catchment Councils in collaboration with ZINWA were required to develop Outline Plans for the purpose of ensuring the optimum development and utilization of the water resources in Zimbabwe (Zimbabwe 1998). The Outline Plans are not exactly IWRM Plans but both have been practiced in Zimbabwe. However, implementation of the 1998 Act has not yielded the desired results (Tom and Munemo 2015) and, as a result, a new Water Policy was formulated in 2013.

The 1994 National Water Policy of Zambia emphasized on a holistic approach to water management in which a comprehensive spectrum of demands are recognized and evaluated to assess their priority. The policy promoted an integrated management approach on water resources development and the use of Helsinki rules which embrace the concept of best joint utilization for the case of internationally shared water resources (Government of the Republic of Zambia [GoZa] 1994). The Water Resources Management Act of 2011 states the principles of water resources management as integrated and sustainable manner with equitable access by all. The Environment is categorized as a water user that shall enjoy second priority of allocation use to the basic human needs. The act provided for the establishment of the Water Resources Management Authority (WRMA) whose function is to promote and adopt a dynamic, gender-sensitive, integrated, interactive, participatory and multi-sectoral approach to water resources management and development. The attributes of the functions of the WRMA have all the hallmarks of IWRM principles although never specifically referred to as such. The act also provided for the establishment of Catchment Councils, Sub-catchment Councils and Water User Associations with the aim of decentralizing water resources management and development functions and involving water users at low levels in decision making. The primary duty of Catchment and Sub-catchment Councils is licensing for water use, among other functions, but also preparing Catchment Management Plans in collaboration with the Water Resources Management Authority for the purpose of ensuring optimum management, development and utilization of Zambia's water resources (Government of the Republic of Zambia 2011).

Conclusion

The brief review of the policies and legal frameworks of the four countries where the detailed description of the hydrology and trend analysis of the catchments were undertaken, reveals some differences in the policies and legal frameworks and also a misstep with the Regional Water Policy of SADC by all countries, save for Tanzania. The regional policy embraces IWRM principles and approaches as best practices for water resources management and development but the policies of most countries and their legislations only allude to some of IWRM attributes, without specifically referring to it, which leaves a lot to be desired in the actual implementation. The extreme variability of climate, anticipated climate change impacts and, consequently, the availability of water in the Zambezi Basin both temporarily and spatially, coupled with the transboundary nature of most sub-basins, calls for an IWRM approach which is practiced from the regional level (ZAMCOM) and reflected at the Zambezi Sub-basin, National and Catchment levels. One would argue that a Zambezi IWRM Plan needs to be developed first, which will then be unpacked into Sub basin, National and Catchment IWRM plans.

The other way round is also feasible: where a regional framework for the development of IWRM plans is drawn such that plans are developed first from the Catchments, National, Sub-basins in that order and, finally, a harmonized Zambezi Basin IWRM Plan is formulated thereof. However, none of these seems to be happening as each country has embarked on its own plans for the utilization, management and development of water resources in their territories, with little attention to the regional body and efforts that have taken nearly three decades to be effectively embraced.

Generally, it was observed that there is a decreasing trend of river flows that are exacerbated by climate change impacts coupled with the projected future increase in water uses that will consequently impact heavily on flow-dependent ecosystem structure and functioning, thereby affecting millions of livelihoods dependent on their services and goods. This situation can only be minimized through a seriously coordinated approach involving all riparian countries with due attention and serious commitment and not the solo approach that is still very much the norm in the capitals of the Zambezi Basin countries.

References

Aasand, F.I., Kammerud, T.A. & Trolldalen, J.M., 1996, 'Challenges in Management of Shared Water Basins: Zambesi River Basin in Southern Africa and Implementation of ZACPLAN', Norway: Resource Geography Group, Department of Geography, University of Oslo.

Beilfuss, R., 2012, 'A Risky Climate for Southern African Hydro-Assessing Hydrological Risks and Consequences for Zambezi River Basin Dams', USA, Berkeley.

Beilfuss, R. & Dos Santos, D., 2001, 'Patterns of Hydrological Change in the Zambezi Delta, Mozambique', Working Paper #2 Program for the Sustainable Management of Cahora Bassa Dam and the Lower Zambezi Valley. Mozambique.

Buishand, T. A., 1984, 'Tests for detecting a shift in the mean of hydrological time series', *Journal of Hydrology*, 73, 51-69.

Dams, J., Dujardin, J., Reggers, R., Bashir, I., Canters, F. & Batelaan, O. J., 2013, 'Mapping impervious surface change from remote sensing for hydrological modeling', *Journal of Hydrology*, 485, 84-95.

DHI, 2015, 'Shire Basin Institutional Planning and DSS: Groundwater assessment report', NIRAS Technical report for the Shire River Basin Programme, Malawi.

Euroconsult Mott Macdonald, 2007, 'Rapid Assessment-Final Report for Integrated Water Resources Management Strategy for the Zambezi River Basin', Lusaka, Zambia: SADC-WD/Zambezi River Authority, SIDA, DANIDA, 206 pp.

Euroconsult Mott Macdonald, 2008, 'Integrated Water Resources Management and Implementation Plan for the Zambezi River Basin', Lusaka, Zambia: Zambezi River Authority (ZRA).

Faraji, S. A., 2016, 'Country Report on the State of Environment and Outlook Report for Lake Malawi/Nyasa/Niassa. A Report For IUCN for the Project on Building River Dialogue and Governance Phase 3 (Bridge-3): 2016-2018 (First Draft)'.

Furusho, C., Chancibault, K. & Andrieu, H., 2013, 'Adapting the coupled hydrological model ISBA-TOPMODEL to the long-term hydrological cycles of suburban rivers: Evaluation and sensitivity analysis', *Journal of Hydrology*, 485, 139-147.

Gerrits, A. M. J., 2005, 'Hydrological modelling of the Zambezi Catchments for gravity measurements', Master's degree dissertation, The Netherlands: Delft University of Science and Technology.

Government of the Republic of Malawi (GoM), 2013, 'Shire River Basin Management Programme (Phase I) Project', Final Environmental and Social Assessment Report, Lilongwe, Malawi: Ministry of Water and Irrigation Development.

Government of The Republic of Zambia (GoZA), 1994, 'National Water Policy'. Lusaka, Zambia: Government of Zambia.

Government of The Republic of Zambia (GoZA), 2011, The Water Resources Management Act No. 21 of 2011, Lusaka, Zambia: Government of Zambia.

Gumindoga, W., Rientjes, T., Shekede, M. D., Rwasoka, D. T., Nhapi, I. & Haile, A. T., 2014, 'Hydrological impacts of urbanization of two catchments in Harare, Zimbabwe', *Remote Sensing*, 6, 12544-12574.

IPCC, 1998, The regional impacts of climate change: An assessment of vulnerability, London, UK: Cambridge University Press.

IPCC, 2007, Climate Change 2007: The physical science basis, Cambridge: Cambridge University Press.

Kaaya, L. T. & Lugomela, G. V., 2015, 'Aquatic Ecosystem Services and Management in East Africa: The Tanzania Case', in L. Chicharo, F. Müller, & N. Fohrer, eds, Ecosystem Services and River Basin Ecohydrology, Netherlands: Springer.

Lahmeyer International, 2013, 'Detailed Design and Investment Preparation for the for the Songwe River Basin Development Programme', In Lahmeyer International Group, ed., A joint project between the Governments of the Republic of Malawi and the United Republic of Tanzania-Update of the 2003 Feasibility Study: Main Report.

Laraque, A., Mahe, G., Orange, D. & Marieu, B., 2001, 'Spatio-temporal variations in hydrological regimes within Central Africa during the XXth Century', *Journal of Hydrology*, 245, 104-117.

Latham, C. J. K., 2001, 'Manyame Catcment Council: A Review of The Reform of the water sector in Zimbabwe', Integrated Water Resources Management: Theory, Practice, Cases, Capetown, pp.148-159.

Lee, A. F. S. & Heghinian, S. M., 1977, 'A shift of the mean level in a sequence of independent normal random variables: a Bayesian approach', *Technometrics*, 19, 503-506.

Macdonald, A. M. & Davies, J., 2000, 'A brief review of groundwater for rural water supply in Sub-Saharan Africa', Technical Report WC/00/33. UK, BGS Keyworth.

Makhanga, S. V., 2011, Contribution of hydrological processes in the occurrence of extreme hydrological events in the Middle Zambezi River Basin. Master's dissertation, University of Zimbabwe.

Mazvimavi, D., Madamombe, E. & Makurira, H., 2007, 'Assessment of environmental flow requirements for river basin planning in Zimbabwe', *Physics and Chemistry of the Earth, Parts A/B/C*, 32, 995-1006.

Mazvimavi, D. & Wolski, P., 2006, 'Long-term variations of annual flows of the Okavango and Zambezi Rivers', Physics and Chemistry of the Earth, Parts A/B/C, 31, 944-951.

Mccartney, M., Cai, X. & Smakhtin, V., 2013, 'Evaluation of the flow regulating function of natural ecosystems in the Zambezi River Basin', IWMI Research Report 148. Colombo, Srilanka: International Water Management Institution (IWMI).

Motsi, K. E., Mangwayana, E. & Giller, K. E., 2002, 'Conflicts and Problems with Water Quality in the Upper Catchment of the Manyame River, Zimbabwe', in P. M. Haygarth, & S. C. Jarvis, eds, Agriculture, Hydrology and Water Quality. Wallingford, UK: CABI Publishing/CAB International.

Mukosa, C., Pitchen, G. & Cadou, C., 1995, 'Recent hydrological trends in the Upper Zambezi and Kafue Basins', in T. Matiza, S. Crafter, & P. Dale, eds, Water resources use in the Zambezi Basin. Proceedings of a workshop held at Kasane, Botswana. Kasane, Botswana: Gland, IUCN.

Mwedzi, T., Bere, T. and Mangadze, T., 2016a, 'Macroinvertebrate assemblages in agricultural, mining, and urban tropical streams: implications for conservation and management', *Environmental Science and Pollution Research*, 1-12 pp.

Mwedzi, T., Katiyo, L., Mugabe, F.T., Bere, T., Bangira, C., Mangadze, T. & Kupika, O.L., 2016b, 'A Spatial Assessment of Stream-Flow Characteristics and hydrologic alterations, post dam construction in the Manyame catchment, Zimbabwe', *Water SA*, 42(2), 194-202.

Nhapi, I., 2009, 'The water situation in Harare, Zimbabwe: a policy and management problem', Water Policy, 11(2), 221-235. Newsday, 2011, 'Is Manyame the most polluted river?' Alpha Media Holdings (AMH), NewsDay, 10 April 2011.

Nicholson, S. E., Klotter, D. & Chavula, G., 2013, 'A detailed rainfall climatology for Malawi, Southern Africa', *International Journal of Climatology*, 34, 315-325.

Pettit, A. N., 1979, 'A Non-Parametric Approach To The Change-Point Problem', *Applied statistics*, 28, 126-135.

SADC, 2004, 'Agreement on the Establishment of the Zambezi Watercourse Commission. Kasane, Botswana', Viewed on 27 July 2016, http://zambezicommission.org/newsite/wp_content/uploads/ZAMCOM%20agreement.pdf.

SADC, 2005, 'Regional Water Policy', Viewed on 27 July 2016, http://www.sadc.int/files/1913/5292/8376/Regional Water Policy.pdf.

Sharma, T. C. & Nyumbu, I. L., 1985, Some hydrological Characteristics of the Upper
Zambezi Basin, Lusaka, Zambia: Kafue Basin Research Committee of the University
of Zambia. Shela, O. N., 2000a, 'Management of shared river basins: The case of the
Zambezi River', *Water Policy*, 2, 65-81.

Shela, O. N., 2000b, 'Naturalization of Lake Malawi level and Shire River flows-Chal-
lenges of water resources research and sustainable utilization of the Lake Malawi-Shire
River System', Paper presented at 1st WARFSA/WATERNET Symposium: Sustainable Use of Water Resources. Maputo.

Shongwe, M. E., Van Oldenborgh, G. J., Van Den Hurk, B. J. J. M., De Boer, B., Coelho, C. A. S. & Van Aalst, M. K., 2009, 'Projected changes in mean and extreme
precipitation in Africa under global warming', *Journal of Climate*, 22, 3819-3837.

SMEC, 2013a, 'Independent Environmental Impact Assessment (ESIA) for the upgraded
Kamuzu Barrage, Final ESIA Vol 1, Main Report, Lilongwe, Malawi: Ministry of
Water Development and Irrigation.

SMEC, 2013b, 'Preparation of an integrated water resources management and development plan
for the Lake Nyasa Basin: Hydrology Report', URT. Tanzania: Ministry of Water (MoW).

SMEC, 2015a, 'Preparation of an Integrated Water Resources Management and Development Plan for Lake Nyasa Basin-Final Climate Change Report', Dar es Salaam,
Tanzania: Ministry of Water (MoW).

SMEC, 2015b, 'Preparation of an Integrated Water Resources Management and Development Plan for Lake Nyasa Basin-Final Hydrology Report', Dar es Salaam, Tanzania: Ministry of Water (MoW).

SMEC, 2015c, 'Preparation of an Integrated Water Resources Management and Development Plan for Lake Nyasa Basin-Final Water Resources Assessment Report', Dar es
Salaam, Tanzania: Ministry of Water (MoW).

Timberlake, J., 2000, 'Biodiversity of the Zambezi Basin', Bulawayo, Zimbabwe: Biodiversity Foundation for Africa.

Tom, T. & Munemo, E., 2015, 'Republic of Zimbabwe National Water Policy: A desk
review of the gaps between the policy and its implementation', *International Journal
of Public Policy and Administration Research*, 2, 60-72.

United Republic of Tanzania (URT), 2002, 'National Water Policy. Dar es Salaam, Tanzania: Ministry of Water and Livestock Development'.

United Republic of Tanzania (URT), 2009, Water Resources Management Act. Dar es
Salaam, Tanzania, Ministry of Water.

Valimba, P., 2004, 'Rainfall variability in southern Africa: Its influences on stream-flow
variations and its relationships with climatic variations', PhD Thesis, Rhodes University.

World Bank. 2006. Zambezi River Basin-Sustainable water resources development for
irrigated agriculture. TFESSD Africa Poverty and Environment Programme, Washington, D.C: The World Bank.

World Bank, 2010, The Zambezi River Basin. A multi-sector investment opportunities
analysis, Washington, D.C: The World Bank.

ZAMCOM, SADC & SARDC, 2015, Zambezi Environment Outlook 2015, Harare,
Zimbabwe, Gaborone, Botswana: ZAMCOM, SADC & SARDC.

ZINWA Act 1998. Zimbabwe National Water Authority Act (Chapter 20:25), Harare,
Zimbabwe: Government Publishers.

7

Environmental Flow Analysis of the Zambezi River Basin

Rashid Tamatamah and Tongayi Mwedzi

Introduction and Definitions

The Brisbane Declaration (2007) provides a useful description of environmental flows: "Environmental flows describe the quantity, quality and timing of water-flows required to sustain freshwater and estuarine ecosystems and the human livelihoods and wellbeing that depend on these ecosystems". This definition emphasizes the linked variables of quantity, quality and timing that together constitute an environmental flow regime of sufficient quality to meet management objectives. For instance, these objectives may be directed at the maintenance or enhancement of the various aquatic and riparian biota and components of the entire riverine ecosystem; at maximizing the production of commercial fish species; at conserving particular endangered species; or protecting features of scientific, cultural or recreational value.

The purpose of an Environmental Flow Assessment (EFA) is to investigate and describe the environmental consequences of modified flow regimes in different parts of a river system, so that the required water quantities can be negotiated along with the required quantities for consumptive uses such as agriculture, livestock, fisheries, wildlife, industry, energy, recreation and other social and economic activities. Typically, EFAs are performed for river systems that are already regulated or are the focus of proposed water resource developments. More recently, however, attention has also been directed at the flow-related aspects of river restoration (e.g. Arthington et al. 2000).

The degree of *'good health'* at which the river will be sustained vary from country to country and region to region depending on a societal judgment about the state in which an ecosystem should be maintained. The required flows will depend on what people want from a river or on the condition at which the river should be maintained. Those values will determine the decisions about how to balance environmental, economic and social aspirations and the uses of the river's waters. Ecological management classes (EMC) (King et al. 2008) is one of the useful concepts employed in balancing ecosystem conservation and water resources development enhancement by providing different flow thresholds corresponding to various river health status and related to different levels of ecosystem integrity maintenance.

Countries all over the world, including those in the Zambezi River Basin (ZRB), have taken up the challenge of determining and implementing environmental flows, and they are written into the water policies of many. One of the overall objectives of these policies is to address cross-sectoral interests in water, watershed management and participatory integrated approaches in water resources planning, development and management. The approaches to conducting environmental flow assessments are many and varied in the level of detail of output, and in data time and resources requirements. Some of the methods are fairly simple and straightforward, but others operate at a complex level; and different methods may be more suitable for particular circumstances.

This chapter describes the importance of environmental flow and provides an overview of the different types of methods and approaches used for determining environmental flow in the Zambezi River Basin. It is not intended to describe individual methods in detail, as many reviews, case studies and manuals on this topic are available elsewhere (inter alia Bovee 1982; Milhous et al. 1989; Arthington and Pusey 1993; Arthington and Zalucki 1998; Tharme 1996; Dunbar et al. 1998; King et al. 1999, 2003; Tharme 2003). The chapter further examines the strengths and limitations of environmental flow assessments carried out in the Zambezi basin, policies and legal framework relating to EFAs in the riparian countries, and provides a brief review of the lessons learned which can be used to improve future EFA undertakings.

Ecological Importance of Environmental Flows in the Zambezi River Basin

A vast body of scientific research recognizes natural flow regime as central to sustaining biodiversity and ecosystem integrity (Poff and Ward 1989; Richter et al. 1997; Rapport et al. 1998; Rosenberg et al. 2000; Arthington et al. 2010). Flow regime is one of the key drivers of the character of a river ecosystem that influence ecosystem components such as channel type and patterns, water chemistry and temperature, bank and associated wetlands around which the native flora and fauna develop (Poff et al. 1997; Arthington et al. 2004). At its most extreme, the long-term absence of environmental flows puts at risk the very existence of dependent ecosystems, and therefore the lives, livelihood and security of downstream communities and industries. Rivers, watersheds, and aquatic ecosystems have long been considered as the biological engines of the planet (Manatunge et al. 2008). They are the basis for life and the livelihoods of local communities by providing a great variety of benefits as highlighted in the works of Costanza et al. (1997) and Postel and Capenter (1997). In Africa, the concentration of rural households along rivers clearly demonstrates the importance of rivers as resources for the sustenance of local livelihoods.

With respect to fisheries, for example, Welcomme (1975) showed strong positive correlation between catches from flood plains on the Kafue River, the Niger Central Delta, and the Lower Shire flood plains and flood levels in the previous year, and indicated that good floods lead to high recruitment. Similarly, a positive relationship between river level and catch was reported in the work of Tweddle et al. (1995) in the Lower Shire. The study by Chimatiro (2004) on the Lower Shire also pointed to the need to provide flood pulses mimicking the natural regime. Chimatiro (2004) modelled the relationships between fish biology and hydrodynamic factors and statistically demonstrated that the flood pulse was the driving force behind major biological cycles of the fish. He also found that the most important measure for increasing yield is the retention of the maximum possible water level during the dry season. It is likely that high floods are directly related to good catches because the extra flooded area results in better juvenile survival and growth of flood plain-loving species such as *Clarias gariepinus*. This suggests that creating smaller floods in some years will be detrimental to fish recruitment (Moyo 1994; Tweddle et al. 1995; Chimatiro 2004).

The impacts of decreasing flows due to rising water demand and long—term river regulation on aquatic ecosystems are becoming increasingly evident in the Zambezi River Basin (Cohen 2013). In the last few decades, the countries of the Zambezi Basin have witnessed rapid urbanization, industrialization, and intensification of agriculture-which have affected the rivers in different ways (Beilfuss et al. 2000; ZAMCOM et al. 2015). The various National Development Plans project significant economic growth. The aspiration in the concerned national vision documents for, among others, Zambia, Tanzania, Mozambique and Malawi (e.g. URT 2010) is to reach Middle-Income Status by 2025-2030. Part of the growth is expected to be derived from industrial development with concomitant growth in industrial water use.

As a result, many rivers are at present highly regulated (Sushka and Napica 1986, Gammelsrød 1992; Mott MacDonald 2007) by multi-purpose reservoirs for water supply, irrigation, hydropower and fisheries. Many flood plains have been cut out from rivers by embankments (Turner 1984; Davies et al. 2000; Beilfuss 2001) and some riparian lands are under intensive agriculture and grazing pressure (Denconsult 1998; Beilfuss and Brown 2006; Nindi 2007; Beilfuss and Brown 2010). Human settlements, deforestation, mining and other activities have degraded the river catchments and increased sediment loads in rivers (Kunz et al. 2011; ZEMA and UNEP 2012). At the same time, some rivers have received increasingly large discharges of industrial effluents, fertilizers and pesticides from agricultural practices and domestic wastes (Chenje et al. 2000; Zurbrügg et al. 2012; ZAMCOM et al. 2015), all of which have affected riverine biota.

Species composition has changed and many species have nearly disappeared (Isaacman and Sneddon 2000; Kamweneshe et al. 2002). The loss of feeding and breeding habitats in the flood plain water bodies due to the construction of embankments, increased silt load and macrophytic growth are major causes for declining fish resources (Tweddle et al. 1995; Hoguane 2000; Chimatiro 2004). The reduction of the supply of sediment to the lower reaches of the Zambezi River by Cabora Bassa Dam may also be responsible for the present marked die-off of the coastal mangroves, and the catastrophic decline of the coastal shrimp-fishing industry (Da Silva 1986; Gammelsrød 1992) (See Box 7.1).

Box 7.1: Some of the ecological effects of river/flow regulations in the Zambezi Basin

The Zambezi River Basin is home to some forty million people. Kariba and Cabora Bassa Dams in the middle Zambezi, and the Itezhi tezhi and Kafue Gorge Dams on the Kafue are among the major dams constructed in the basin. The urgent need to provide more water services within the Zambezi River Basin has conflicted with the desire to maintain or improve the ecological condition of these rivers.

Lake Kariba and Cabora Bassa

These impoundments have had a major impact on the economy of the region as they represent an important source of hydroelectricity for the region. Fish diversity is high in the Zambezi Basin, with a best estimate of 165 different freshwater species in the Zambezi basin in addition to more than 500 fish species in Lake Malawi/ Nyasa (Chenje 2000). The Lake Tanganyika sardine *Limnothrissa miodon* (colloquially known as 'kapenta') which was introduced into Lake Kariba, but later survived discharge via the Kariba turbines to colonize Cabora Bassa, has formed the basis of an important fishing industry in the basin (Davies 1986). Nevertheless, these economic benefits have come at considerable ecological costs. The dams have affected major flood plains due to reduction of the supply of water and sediment. The resultant contractions of these wetlands have impacted fish, avifauna and water-dependent antelope species (Davies 1986).

Kafue Flats

The Kafue Flats in Zambia, an area of 4380-7000 km^2 of flood plain on the Kafue Rivers was once the most productive fishery in Zambia, and included some of the most spectacular concentrations and vulnerable species of mammals and birds in Africa (Bingham 1978). In addition, flood plain grazing supported some 700,000 cattle, a quarter of the national herd (Ghirotti et al. 1991). With the construction of the Kafue Gorge Dam in 1971 and its regulating reservoir at Itezhi-tezhi in 1978, river regulation was instituted. Since 1972, regulation has attenuated the flood peaks and raised base flows; drainage has been retarded, the area of flood plain reduced and unseasonable fluctuations in water level induced. The construction of the Kafue dams has led to decline in fish production and fish biodiversity, and decline in flood plain pasture.

The Lower Zambezi and Zambezi Delta

In the Lower Zambezi and Zambezi Delta in Mozambique, river regulation began with the closure of the dam at Kariba in 1958, but with the completion of Cabora Bassa in 1975, flows in the lower Zambezi dramatically changed. There is now little seasonal variation in river flow at Tete. The natural flood has been attenuated and the base flow increased; flood peaks are unpredictable and may occur at any time. It has also been estimated that the closure of Cabora Bassa led to a 70 per cent reduction of the supply of sediment to the lower reaches of the river, and that this has led to severe erosion of the Zambezi Delta, threatening both agricultural activity and the important Marromeu wilderness area. It may also be responsible for the present marked die-off of the coastal mangroves, and the catastrophic decline of the coastal shrimp-fishing industry around the delta (Davies 1986). Furthermore, during the annual floods prior to construction of the dams, the drowned grasslands of the delta were protected from overgrazing, and thus able to regenerate. The virtual elimination of the natural flooding regime has disturbed this important ecological dynamic.

Source: Davis et al. (2000)

Methods of Determining Environmental Flows

A global review of the present status of environmental flow methodologies (Tharme 2003) revealed the existence of some 207 individual methodologies, recorded for 44 countries within six world regions. The methodologies range from relatively simplistic, reconnaissance-level approaches for the early phases of water resource planning initiatives, to resource-intensive methodologies for highly modified, individual catchments or sites.

Methodologies for environmental flow assessment (EFA) vary in levels of data requirements and complexity, and the majority fall into one of four general categories: (1) hydrology-based methodologies; (2) hydraulics-related methodologies; (3) habitat simulation methodologies; and (4) holistic methodologies, or hybrid methodologies derived from components of methodologies that fall in these different categories (Tharme 2003). Differences in group classifications have also been reported among authors (e.g., Acreman and Dunbar 2004; Jacimovic and O'Keeffe 2008). Table 7.1 provides a general overview of EFA methods, many of which have been employed in different parts of the Zambezi River Basin.

Table 7.1: An overview of seven generic types of Environmental Flow Assessment

Type	Approach	Required data	Required time	Estimation of funds required	Specialist expertise required	Advantages and constraints
Hydrology–based	Look-up table (e.g. Tennant)	Existing or modelled flow data	1 day	< $ 5 000	Some hydrological knowledge, and ecological insight	Low confidence, general results, low costs, quick.
	Hydrology–based (e.g. IHA)	Existing or modelled flow data	1 day–1 months	< $ 10 000	–Some hydrological knowledge; –Ecological insight	Low confidence, general results, low costs, quick.
	Extrapolation (e.g. Hughes Desktop)	Based on correlation with existing detailed studies	1 day	$200 000 to develop, < $ 10 000 to apply	–Hydrologist, –Modeller, –Ecological insight	Possible for regions having numerous assessments through comprehensive methods, to provide the dataset for extrapolation, low confidence, low costs, quick.
Hydraulic rating	Hydraulic rating	Surveyed cross-sections	≤ 3 months	< $ 50 000	–Hydraulics engineer, –Hydrologist, –Ecological insight	No/few ecological inputs, low/medium confidence.
Habitat simulation	Habitat simulation (e.g. IFIM)	Hydraulic habitat requirements of target species, Multiple–rated hydraulic cross–sections	3 months–1 year	$ 250 000–$3 million	–Hydraulics engineer, –Biologist, –Hydrologist, –Modeller	High confidence for target species, but lacks ecosystem focus.
Holistic	Site–based Comprehensive (e.g. BBM, DRIFT)	• Existing and sampled; • Biophysical and social data; • Hydraulic cross sections; • Socio	1–3 years	$ 150 000–$3 million	–Hydrologist, –Hydraulics engineer, –Freshwater biologists, –Geomorphologist, –Water quality specialist, –Socio–economist	Ecosystem based, high confidence, socio–economic factors included, high costs of resources
	Regional (e.g. ELOHA, PROBFLO)	• Regional hydrologic; • classification; • Flow–ecology; • Relationships	1 year to develop; 1–3 months to Apply	$ 1 million to develop; $100 000 to apply	–Hydrologist; –Hydraulics engineer; –Freshwater biologist:–Geomorphologist; –Water quality specialist; –Socio–economist	Based on regional ecology and hydrology, high confidence, socio–economic factors included.

Modified from Jacimovic and O'Keeffe 2008

Environmental Flow Studies/Assessments in the Zambezi Basin

This review on the present status of environmental flow assessment revealed the existence of approximately 25 individual environmental flow assessments conducted in the countries of the Zambezi Basin both on rivers within and outside the basin. As is the tendency with any new creative science, the approaches to doing this are many and varied, some successful and others less successful, some operating at a minimalist and others at a complex level.

The EFA methodologies employed in the assessments conducted in the basin are typically of two levels.

- ▶ The majority were at reconnaissance-level initiatives, relying on hydrological methods namely, desktop reserve model widely used in Malawi and Tanzania; the range of variability approach (RVA) used in the Manyame catchment, Zimbabwe; and the modified Tennant method.
- ▶ At more comprehensive scales of assessment, a small number of these assessments have used modified versions of holistic methodologies.

Some of the methods used are scenario-based, providing not only explicit links between changes in flow regime and the consequences for the biophysical environment, but also consideration of ecosystem-dependent livelihoods and a benchmarking process suitable for evaluating alternative water resource developments at basin scale, in relatively poorly known systems (e.g., DRIFT-based EFAs conducted in the Pangani Basin, Tanzania, and the mouth of Zambezi Delta in Mozambique). Table 7.2 presents results of some of the EFA studies undertaken in the Zambezi Basin rivers/lakes; and EFAs conducted in the Manyame catchment in Zimbabwe, Great Ruaha catchment in Tanzania and 20 sub-basins in Malawi are presented in case studies 1, 2 and 3 respectively.

The results from the EFAs carried out in the basin indicate that during years of normal rainfall (see Chapter 6 of this volume), there is ample flow in rivers to meet the environmental flow recommendations and allow for significant extractive uses. However, the flow available for abstraction is concentrated during wet months and far less water is available for abstraction during dry months (e.g., Scott 2003; GoM 2010; WWF-TCO 2010). This calls for the need to incorporate environmentally-sensitive infrastructure to store water in the wet season for use during dry months.

The need to sustain small flows during the most severe droughts also highlights the importance of interventions in the upper basin to manage land to maximize infiltration and groundwater recharge. Current patterns of deforestation; overgrazing and small-scale farming practices tend to reduce infiltration while increasing overland flow (EPSMO-BIOKAVANGO 2009; WWF-TCO 2010). Governmental and non-governmental organizations are promoting improved land use practices in the headwater catchments and some are considering innovative payments for watershed services to finance them (Turpie et al. 1999; Scholes and Biggs 2004; McCartney et al. 2013). They argue that higher water prices could encourage conservation (thus increasing river flows).

It is acknowledged that while there is some uncertainty in the estimates for environmental flow requirements in rivers across the basin, these numbers are the only quantitative values currently available from structured and scientifically sound processes. The estimates should be refined and adapted as more information becomes available in the sub-basins. The next step will be to translate these flow recommendations into policies and practices on the ground that can guide the sustainable development of water resources into the future, while protecting basic water needs for people and the ecosystems upon which they depend. For example, environmental flow recommendations for transboundary Mara River were approved by the Council of Ministers of the East African Community (GLOWS-FIU 2012), which has called upon Kenyan and Tanzanian water management agencies to take steps to implement them. Similarly, the Southern African Development Community (SADC), of which many riparian countries in the Zambezi Basin are members, adopted a SADC Water Policy that calls upon member countries to allocate adequate water for maintaining ecosystem integrity. The Revised SADC Protocol on Shared Watercourse Systems (SADC 2000) requires that member countries should aim to achieve a balance between water development and protection of the environment.

Table 7.2: Environmental Flow Assessments in different parts of the Zambezi River Basin

Country	River system	EFR Data	EFA Method	Source
Angola	No data available on rivers within the Zambezi basin	Nil	-	-
	Data for other basins	Nata River: 2.9 m^3s^{-1} Shashe River: 6 m^3s^{-1}	Hydrological modelling	Epsmo Biokavango (2009)
Botswana	Okavango	390 m^3s^{-1}	Hydrological modelling	King et al. (2009)
Malawi	All rivers	A minimum dry and wet season flows expressed as a percentage of mean annual flow (nMAR).	Hydrological modelling	GoM (2015)

Mozambique	Marromeu Complex of the Zambezi Delta	A two-week release in February to generate flows of 4,500 m³s⁻¹	DRIFT	Beilfuss et al. (2006)
	Buzi River	To maintain the ecological status in the Buzi River at a largely natural condition (ecological category A) an average allocation of 57 percent of mean annual runoff (nMAR) is required.	Desktop	Lagerblad (2011)
Namibia	Lake Liambezi	Maintain Elev. 935.8 m at the intake sill which occurs when discharge in the Cuando/ Linyanti/ Chobe/ River exceeds 2,500-3,000 m³s⁻¹.	Desktop	Hughes (2004)
	Rivers within the Zambezi Basin (L. Nyasa catchment)	Forty per cent of the monthly median flows, which in 2015 is equivalent to the following EFRs (Mm³year⁻¹) for individual sub-basins: Songwe-412, Kiwira-605, Mbaka-338, Lufirio-381, Rumakali-288, Lumbira-410, Nkiwe-260, Ruhuhu-1838, Mchuchuma-54, Mbawa-232	Tennant	SMEC (2013)
Tanzania	Data for other basins	Great Ruaha-WWF project Dry season (Oct) flow at Msembe Ferry (Ruaha Nat. Park)-303.2 Mm³ (13.8 per cent of Present MAR) = 0.5-1.0 m³s⁻¹	BBM	WWF-TCO (2010)
		Ruvuma River (Mm³): upper 1108.7, middle 622.5, lower 225.5	Desktop	Atkins & Humphreys (2013)
		Wami e.g. Matipwili (most downstream EFA site) Driest year: Dry season (October=4.6 m³s⁻¹; Wet season (April)=21.2 m³s⁻¹ Maintenance year: Dry season (October)=6.6 m³s⁻¹, Wet season (April)=39.0 m³s⁻¹ Wettest year:Dry season (October)-37 m³s⁻¹, Wet season (April)-235 m³s⁻¹	A hybrid of Savannah and BBM	GLOWS WRBWB (2013)
		Ruvu-e.g. at Kongo (most down-stream EFA site) Dry year : Dry season (November)=4.9 m³s⁻¹ Wet season (May)= 22.00 m³s⁻¹ Maintenance year Dry season (October)= 7.5 m³s⁻¹ Wet season (April) =67.7 m³s⁻¹	A hybrid of Savannah and BBM	GLOWS WRBWB (2013)

		Mara-e.g. at Kogatende (EFA Site 4-Serengeti Nat. Park) Dry year: Dry season (November)=1.81 m^3s^{-1} Wet season: (May)=5.00 m^3s^{-1} Maintenance year: Dry season (October)=8.85 m^3s^{-1} Wet season (April)=23.50 m^3s^{-1}	BBM	GLOWS WRBWB (2013)
		Songwe at Galula (EFA 2) Dry year: Dry season (October)=1.5 m^3s^{-1} Wet season (March)=40.0 m^3s^{-1} Maintenance year: Dry season (October)=0.2 m^3s^{-1} Wet season (March)=60.0 m^3s^{-1}	BBM	WREM (2014)
		Katumaat Stalike(EFA 1): Dry year: Dry season (October)=0.12 m^3s^{-1} Wet season (April)=15.00 m^3s^{-1} Maintenance year Dry season (October)=0.2 m^3s^{-1} Wet season (April)=30.0 m^3s^{-1}	BBM	WREM (2014)
Zambia	Itezhi-Tezhi dam release	400 m^3/s in February and 600 m^3/s in March and April is required	Desktop	Scott (2003)
Zimbabwe	Manyame River	Class A rivers-41-67.5 mm^3 for upper Manyame and 15-35 mm^3 for lower Manyame. Class B rivers-31-52.5 mm^3 for up-3 lower Manyame. Class C rivers 16-30 mm^3 for upper Manyame and 7-15 mm^3 for lower Manyame. See Table 7.3 for river class	Desktop Reserve Model	Mazvimavi et al. (2007)

Case Study 1: EFA of the Manyame River Catchment, Zimbabwe

The most comprehensive study done on determining the environmental flow of the Manyame Catchment (and the whole country in general) was by Mazvimavi et al. (2007). The study employed the Desktop Reserve Model (DRV) (Hughes and Hannart 2003). The EFR in this model depends on the environmental management class (the desirable target to be maintained) of the particular river section. The DRV uses four environmental management classes developed by O'Keefe and Louw (2000). Class A rivers have unmodified habitat (i.e., they have natural conditions), Class B rivers have few modifications but largely natural conditions, Class C rivers have moderate modifications but with unchanged ecosystems, and Class D rivers have modifications which have caused substantial losses of habitats or degradation. Mazvimavi et al. (2007) deliberately left out Class D rivers, estimating their EFRs on sections that have largely been unaffected by damming and impoundments. In this method (DRV), the Environmental Flow Requirements are given as a proportion of the Mean Annual Runoff (MAR).

Mazvimavi et al. (2007) showed that the EFRs of the Manyame catchment vary as you move from upper Manyame to lower Manyame catchment (Table 7.3). This is expected as the flow regime of the Manyame Catchment is known to vary as well (See Chapter 6 of this volume). The study then went on to explore the current water allocation against the EFRs. Smakhtin et al. (2004) alluded that if the water allocated for use within a catchment is less than 30 per cent of MAR then the EFR is being slightly used. 30-60 per cent allocation of MAR to water use represents moderate; 60-100 per cent heavy and >100 per cent over-utilization of water that should have been allocated as environmental flows. The total amount of water allocated for both storage and abstraction in the Manyame Catchment is shown in Figure 7.1. The figure clearly shows that water that should be reserved as environmental flows is being overexploited in the catchment, with most sections of the catchment being allocated more than 100 per cent of the MAR for utilization.

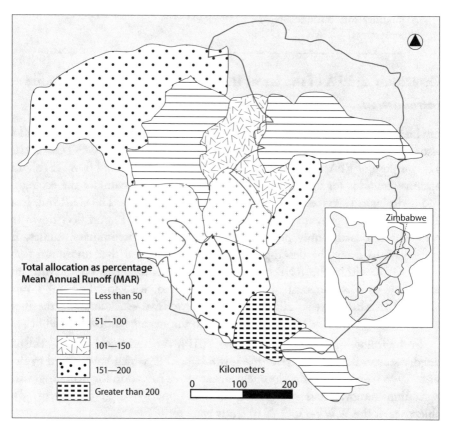

Figure 7.1: Total water allocated for storage and abstraction (as a percentage of Mean Annual Runoff, MAR) in Manyame Catchment, Zimbabwe (adapted from Mazvimavi et al. 2007)

Table 7.3: Environmental Flow requirements of the Manyame Catchment as a proportion of the Mean Annual Runoff (MAR, mm year^{-1})

Environmental Management (target) class	Upper Manyame (MAR of 101-150)	Lower Manyame (MAR of 51-100)
Class A : (Rivers with natural conditions—unmodified habitat)	41-45 per cent	31-35 per cent
Class B: (Rivers with few modifications but largely natural conditions)	26-30 per cent	21-25 per cent
Class C: (Rivers with moderate modifications but unchanged ecosystems)	16-20 per cent	14-15 per cent

Case Study 2: EFA of the Great Ruaha River Catchment, Tanzania

Introduction

Environmental Flow Assessment (EFA) for the Great Ruaha River (GRR) and its associated wetland were conducted in the year 2008-2010 (WWF-TCO 2010). The purpose of EFA was to determine required environmental flows (EFs), and options/scenarios for implementing the EFs in order to maintain the ecological and hydrological services provided by the GRR ecosystems. The GRR rises from Usangu wetland upstream of Ruaha National Parks (RNP) and flow down the park where it is the only perennial water source. In preliminary studies, EF investigations using the desktop reserve model suggested that minimum flows in the GRR within the RNP of 0.6 m^{3}s^{-1} (corresponding to 635.3 Mm^3 and equivalent to 21.6 per cent of MAR) are required, with low flows of 1 m^{3}s^{-1} preferred (Kashaigili et al. 2007). Given the extensive water losses in the Ihefu swamp (Usangu wetland), this corresponds to the swamps inflows of 7-8 m^{3}s^{-1}.

In the follow-up study (WWF-TCO 2010), the main goal was to reinstate the dry season flows in the river reach through RNP which have ceased to flow since 1993 owing to the development of extensive irrigation for rice, progressive vegetation removal and overgrazing in the upper Usangu catchment. The objectives of the WWF-TCO 2010 study were to:

    i.   Recommend flow rates for different seasonal scenarios required to restore the dry season flows to the middle section of the GRR in the Ruaha National Park (RNP);

ii. Identify a range of options to support implementation of environmental flows, providing a short-list of preferred options identified against agreed criteria in consultation with a wide range of stakeholders at local and national levels in Tanzania;

iii. Determine the required inflows into the eastern Usangu wetland, in order to meet the recommended flow rates downstream and in the RNP;

iv. Determine the response of the wetland to changing flow regimes, not only those caused by upstream abstraction, but also with respect to proposed engineering modifications, i.e. the construction of the Lugoda Dam, and the Ndembera transfer option.

The BBM method was used for EFA at two representative sites within RNP, BBM1 at Msembe and BBM2 at Muhuwa (Figure 7.2). The EFA then determined the volume, duration and timing of flows required to meet the above objectives.

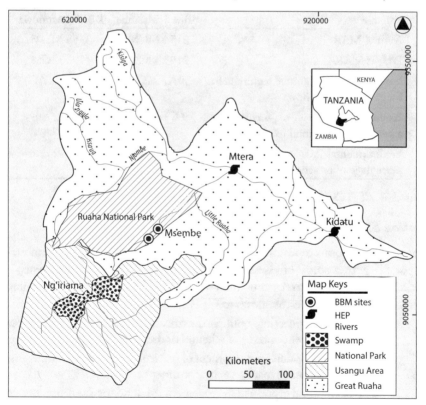

Figure 7.2: The Great Ruaha River Catchment and the location of Building Block Methodology stations (Adapted from WWF–TCO 2010)

Results

EFA results for the two sites in the GRR are summarized in Table 7.4. The confidence rating by the EFA specialists ranged from 3 (Moderate confidence) to 5 (Very high confidence). It was estimated that a total reduction of 65-90 per cent of current dry-season abstraction would be required to achieve the necessary Ihefu inflows to meet the EFR at RNP. Given the economic and social requirements for water in the Usangu Plains and that Tanzania has prioritized irrigation agriculture in her key economic growth and poverty reduction strategies, achieving the suggested inflows by particularly reducing water uses is practically non-feasible. Accordingly, an alternative solution to this problem had to be found.

Table 7.4: Summary of recommended flows in terms of natural and present Mean Annual Runoff (MAR), in Million Cubic Metres (MCM) at Building Block Methodology Stations 1 and 2

Variable	BBM 1 Msembe	BBM 2 Muhuwa
Natural MAR	3154 MCM	3154 MCM
Present MAR	2193 MCM	2193 MCM
Long term average annual requirement for environmental flows	303.2 MCM	324 MCM
Environmental flow requirement as a percentage of natural MAR	9.6 per cent	10.3 per cent
Environmental flow requirement as a percentage of present MAR	13.8per cent	15.0 per cent

Source: WWF-TCO (2010)

Option Study

The study of options to restore flows to the river within the RNP was undertaken in parallel with a study of assessing the environmental flow requirements. The options were subjected to rigorous analysis including the stakeholder consultation workshop. Options to restore flows to GRR:

- ▶ Technical (engineering) options: the construction of infrastructure, storage, and development of transfer schemes or the abstraction of groundwater;
- ▶ Institutional options: development of particular institutional arrangements and water resource management instruments;
- ▶ Environmental options: management of land or land-use to retain environmental services and functions;
- ▶ Agricultural options: management of agricultural activity and adaptation of agricultural practices; and
- ▶ Economic options: linked to economic development and the provision of alternatives to current economic practices.

In addition to stakeholder consultations, the wetland EFA also assessed the extent to which the proposed flow restoration options might have an impact on the river. From the evaluation and consultation process that was conducted, three out of the five options given above were preferred for the restoration of flows in the GRR. They include:

i.   Institutional strengthening and support to ensure improved water resources management, including improved management of irrigation water;

ii. Construction of an impoundment on the Ndembera River (Lugoda Dam);

iii. ransfer of flow from the Ndembera River.

In order to fully restore flows to the GRR, there was need to implement all three options needed concurrently. However, since options 1 and 2 were only likely to achieve this objective within the medium to long term, the third option transfer from the Ndembera River was highlighted as the most likely to restore flows in the short term. An urgent intervention was required in the short term to prevent further degradation of the Great Ruaha River and loss of natural (and financial) capital within the RNP.

For the Ndembera transfer option, the water could be transferred in-channel through the Eastern wetland. Given the existing situation of zero river inflows into Ihefu, the entire 6.81 m^3s^{-1} would need to come from the Ndembera River, which could not be assured without the reservoir. It was determined that any required high flow (> 1 m^3s^{-1}) across the BBM sites would require a much higher discharge ($> 0.93-6.81$ m^3s^{-1}) from the Ndembera River (WWF-TCO 2010). At the time of writing this report, plans were in advanced stages for the construction of Lugoda Dam across Ndembera River.

The following observations were made from this study regarding the implementation of the recommended EFs in order to maintain the ecological and hydrological services provided by the GRR ecosystems:

▶ There is need to improve WRM for the long-term sustainability of any approach to restore flows within the GRR in the short to medium-term.

▶ Without appropriate institutional arrangements and improved management of the water resources of the GRR, any gains achieved in the short term in terms of flow within the GRR will likely be negated through increased abstraction, and unsustainable agricultural and land-use development of the Usangu catchment (highlands and plains).

▶ Careful monitoring of the results of implementing EFs in the GRR will be important, to determine whether the flows are achieving the desired state of the river. Accordingly, some flexibility may be required in water resource management policies and decision making in the catchment. This will also allow for successive modifications in the light of increased knowledge, changing priorities, and changes in infrastructure (e.g., removal of dams) over time.

Case Study 3: EFA in the Malawian Rivers

In Malawi a method similar to desktop reserve model (Hughes and Munster 1999) was used to determine Environmental Flow requirements (EFRs) for 20 sub-basins across the country to inform the Water Resources Investment Strategy (WRIS) (GoM 2011). The method provides descriptions of flow regimes that would maintain regulated river ecosystems in certain catchment conditions (categories A to D, with A being Natural and D being Largely Modified) (Kleynhans 1996).

Areas within Malawi that are hydrologically significantly different from each other were delineated. All twenty points across the country were located directly at flow gauging stations due to the difficulty of estimating flow statistics at ungauged sites. Ideally, at least one point in each of the 17 Water Resource Areas (WRAs) in Malawi would have been chosen but a paucity of reliable data in some WRAs or parts of WRAs prevented the setting of the EFR stations in every WRA. Other criteria for locating EFR points across WRAs included high quality catchments in need of protection, or protected areas such as National Parks, and areas of high social dependence on watercourses.

The hydrological indices (HIs) of the EFR locations in relation to the existing habitat integrity condition A-D were determined. The hydrological index (HI) is an overall index of flow variability based on the combination of the index of the coefficient of variation (CV index) and the base flow index (BFI). The hydrological indices derived for Malawian rivers ranged in value from 0.82 in the north Rumphi catchment (designated as WRA7) to 11.03 in the Dwangwa catchment (designated as WRA6).

$$HI = CV\ index\ /\ BFI$$

Based on the South African Desktop Model, the EFRs required to maintain rivers in a B, C or D catchment (ecological) condition were generated. These EFR requirements were then expressed as a percentage of nMAR so that these percentages could be used to calculate exact EFR requirements for Malawian rivers, based on actual flow data recorded. Where available and applicable the final EFR requirements for Malawian rivers were then augmented by comparison with those from full EFR assessments for South African rivers with HI ranges similar to those generated for the rivers in Malawi.

In consultation with staff in the Ministry of Irrigation and Water Development (MoIWD), and in the Land Resources Conservation Department, the current catchment (ecological) condition (B, C or D) for each chosen EFR point in Malawi was determined using Kleynhans (1996). The hydrological indices in relation to existing habitat integrity are presented in Figure 7.3, and The EFR estimates (as percentage nMAR) are summarized in Table 7.5. This process effectively allocated a volume of water in Mld^{-1} to environmental and social needs for each of the catchments (WRAs) in which the process was implemented.

Table 7.5: Environmental Flow Requirements in relation to existing catchment conditions

WRA	Name	Existing catchment conditions	EFR to be maintained over each period expressed as a percentage (%) of mean annual flow		
			Dry season	Wet season	Annual
1	Shire River	D	15	24	19
2	Lake Chilwa	C	16	33	24
3	South West Lakeshore	C	10	42	26
4	Linthipe	D	10	32	21
5	Bua	C	10	42	26
6	Dwangwa	C	16	33	24
7(A-G)	South Rukuru	C	26	37	31
7(H)	North Rumphi	B	32	69	51
8	North Rukuru	B	32	69	51
9(A)	Lufira	C	26	37	31
9(B)	Songwe	B	32	69	51
10	South East Lakeshore	D	10	32	21
11	Lake Chiuta	D	10	32	21
12	Likoma Island	-	-	-	-
13	Chizumulu Island	-	-	-	-
14	Ruo	C	10	42	26
15	Nkhotakota Lakeshore	C	10	42	26
16	Nkhata Bay Lakeshore	B	32	69	51
17	Karonga Lakeshore	B	32	69	51

Source: GoM (2015)

The volume of water determined as the EFR was then "removed" from the supply volumes in supply-demand calculations as it is not available for other purposes, such as abstraction, irrigation, etc. There were a few caveats identified regarding the calculation of the EFR requirements. They include the following:

▶ The Malawi HIs are based on measured daily flow data, whereas they should be calculated using naturalized data. The implication is that areas could be characterized wrongly because of highly modified flows resulting from water use. This would lead to an incorrect HI being calculated, followed by an incorrect EF allocation.

▶ The characteristics of individual reaches, e.g., location in the catchment (mountain stream, foothill, lower river, etc.), channel shape and key species are not known and were not taken into account in assigning EFs. These characteristics are known to affect the volume and distribution of water required for maintenance.

▶   In some cases in South Africa, full EFs have yielded higher EF estimates than those provided by the Desktop, and so recommended Desktop percentages should be used with caution.

Country Policies and Legal Framework

In the last decade countries in the Zambezi River Basin developed laws and policies that give priority of water to river ecosystems once basic human needs are met. These countries passed new legislations aimed towards ensuring access to safe water resources for all people, as well as sustaining the valuable ecosystems upon which these people depend. The principle of environmental flows is evident in the wording of these laws. Although national water policies of some riparian countries of the Zambezi have been presented in detail in Chapter 6 of this book, we present them here in their relevance to EFF. For example, in Zimbabwe the management of the water resources is implemented based on two acts, the Zimbabwe National Water Authority Act (1996) and the Water Act (1998). By these two Acts, the Zimbabwe National Water Authority (ZINWA) manages water resources on a catchment basis with the involvement of stakeholders (through catchment councils) in each catchment area. In terms of the Water Act of 1998, Part II, section 12, each of the catchment councils has a responsibility of producing a catchment outline plan for every river system. It is in the catchment outline plan that environmental flows are protected. In section 13 of the Water Act of 1998, the outline plan must indicate measures for the conservation and improvement of the physical environment and the proportion of the available water reserved for future use or for the benefit of the environment. Furthermore, section 67 requires water resource management to be consistent with environmental approaches. It requires that due consideration be given to:

> *"(a) the protection, conservation and sustenance of the environment; and (b) the right of access by members of the public to places of leisure or natural beauty related to water or water bodies."*

In Tanzania, the Tanzanian Water Policy (2002) (URT 2002) recognizes that, "…In-stream flows or environmental flows and levels are necessary for riparian biodiversity, wetland systems, freshwater-seawater balance in deltas and estuaries (Section I, 2.8)." Hence, it aims to guarantee "[a] minimum water requirement... to all humans to maintain human health, and sufficient water... to restore and maintain the health, services and functions of ecosystems (Section 3, 3.1 (i))." Water use is prioritized such that "…water for basic human needs in adequate quantity and acceptable quality will receive highest priority. Water for the environment to protect the eco-systems that underpin the water resources, now and in the future will attain second priority and will be reserved" .

Following the National Water Policy in 2002, the Water Resources Management Act (2009) (URT 2009) has given effect to this Policy and also replaces the Water Utilization (Control and Regulation) Act No. 42 of 1974 and its subsequent amendments. The new Act is intended to ensure that the nation's water resources are protected, used, developed, conserved, managed and controlled for basic human needs for the present and future generation. Part VI (a. Water Resources Classification and Reserve) provides for protection of water resources which includes water resources classification and reserve (definition provided in the section above), establishment of protected zones, declaration of groundwater controlled areas and prevention of pollution. This includes that the Minister can Gazette the classification of a resource, specify the resource quality objectives of the class to which the resource belongs, and the requirements for achieving these objectives.

Further, the Environmental Management Act No. 20 of 2004, section 60 (3) places responsibility on Basin Water Boards (BWB) to prioritize different uses of water such that adequate water reserve is available for the environment. Similarly, the Zimbabwe Water Act of 1998 grants the Minister rights to gazette information obtained from research on hydrological matters such as the quantity and quality of water resources, the utilization thereof and how to develop water resources to meet the country's needs.

Discussion

Environmental flows form part of an ecosystem approach to integrated water resources management. By ensuring the continued availability of the many benefits that healthy river and groundwater systems bring to society, environmental flows provide critical contributions to river health, economic development and poverty alleviation. It is increasingly clear that failure to meet environmental flow requirements has disastrous consequences for river users in the Zambezi Basin. However, meeting the water needs of aquatic ecosystems will often mean reducing the water use of one or more sectors. These are tough choices the riparian countries have to make to ensure the long-term health of the basin and the activities encompassed.

In the last fifteen years, more than 25 individual environmental flow assessments have been conducted in the countries of the Zambezi Basin. Of the types conducted, the look-up table type (e.g. Tennant's Montana method) hydrology-based methods) have been frequently applied in the basin and followed by a few occasions in which Holistic methods such as BBM and DRIFT were also used. Regardless of the EFA method used, on average EFs equivalent to 10-30 per

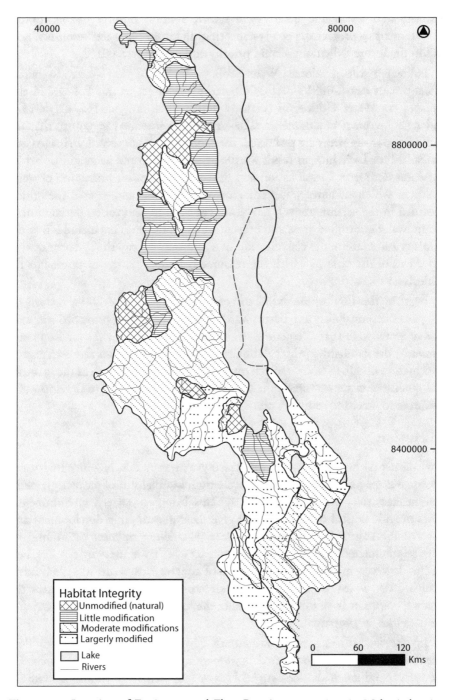

Figure 7.3: Location of Environmental Flow Requirement points in Malawi showing existing habitat integrity scores and Hydrological Indices.

Source: GoM (2015)

cent of MAR have been recommended in rivers across the basin during the dry season and 30-60 per cent during the wet seasons. Individual differences within each category are dependent on the environmental management class prescribed for individual river/site. Important observations made from these assessments is that EFA studies lacked consistent criteria for selection of methodologies used, possibly reflecting the fact that the studies were supported by different donor agencies/NGOs. Also, at present there are no guidelines or regulations in place in the riparian countries to guide the design of these EFA projects. Lack of guidelines delays the much-needed starting point towards actual implementation or incorporation in the water allocation process of the required Reserve as stipulated in Water Resources Management laws of these countries. While this is not a focus of this chapter, it illustrates that these EFA projects have had influence at national coordination levels.

It is important that a critical review of these different methods be conducted so that countries benefit by identifying a small number of methods that are suited to the different circumstances across their river and lake basins and are also cost-effective for riparian nations that face enormous budgetary constraints. Experience from Tanzania shows that different methodologies have been used to estimate EFs during IWRMD plan developments in different basins, varying from simple hydrology-based look-up tables to complex holistic methods. As a result, some rivers draining highly socio-economic important areas earmarked for the country's agricultural and industrial development were subjected to estimations using simple look-up table EFA methods, which have low confidence for conserving aquatic and riparian ecological integrity in areas where anthropogenic activities are intense or predicted to intensify (MoW 2015).

Conclusion

In order to improve the way in which EFAs are conducted and establish systematic procedures for determining the reserve, there is need to develop and adopt guidelines or criteria for conducting EFAs. The overall objective of these guidelines would be to provide adequate tools for establishing water needs of the aquatic and riparian ecosystems for different types of water bodies, which must be incorporated in the water allocation process. In many instances, this should begin with the classification of water bodies aimed at creating homogeneous groups representing water bodies with identical characteristics in relation to the broad objective of classification. Hydrological changes that a water body undergoes normally correspond to volumetric and/or regime changes and affects water bodies differently with respect to their types and geographical location. It is therefore important to classify water bodies according to attributes defining type, size, regime and location.

The guidelines should also provide a range of options at different levels of detail for EFA on water bodies depending on the time and resources available, the level of expertise, data and information available, and the importance, scale, level of development and management issues that pertain to the river under consideration. For example, in the environmental reserve guidelines being developed for Tanzania (MoW 2015), at least two potential levels of assessment detail are proposed for carrying out EFAs. The first level EFA is an appropriate hydrology-based model. This will be applied to all water bodies, so that preliminary estimates of water necessary for the ecological Reserve for any water body are instantly available. This should provide managers with a first-level tool to assess the possibilities of allocating water (either for the Reserve or for additional users) from any water body. The second level EFA would be a holistic/comprehensive methodology such as the BBM, HFSR or DRIFT. Such a method would be necessary in situations where ecosystem integrity has been or anticipated to be compromised in areas where anthropogenic activities are intense or predicted to intensify.

To be relevant, investments in environmental flows will need to be justified by improvements in environmental, social or economic conditions within the wider society, rather than on the basis of the impacts felt by specific actors or donors. Without societal benefits there will be little financial and economic rationale for undertaking and financing the required changes. High priority should be given to those situations where the direct benefits are clear, in particular for the poor, and where the applied methods are cost-effective and well known.

On a positive note, EFA projects have provided opportunities to develop local technical skills for undertaking further EFAs (including for project developments) within the countries where they have been undertaken. Many of the existing EFAs were specifically designed to build these skills across government institutions at the national and basin levels, as well as academic institutions.

References

Acreman, M. & Dunbar, M.J., 2004, 'Defining environmental river flow requirements-a review', *Hydrology and Earth System Sciences*, 8(5), 861-876.

Arthington, A.H. & Pusey B.J., 1993, 'In-stream flow management in Australia: methods, deficiencies and future directions. *Australian Biology*, 6, 52-60.

Arthington, A.H. & Zalucki, J.M., eds, 1998, 'Comparative Evaluation of Environmental Flow Assessment Techniques: Review of Methods', Occasional Paper No. 27/98. Canberra, Australia: Land and Water Resources Research and Development Corporation.

Arthington, A.H., Brizga, S.O, Choy, S.C, Kennard, M.J., Mackay, S.J., Mccosker, R.O., Ruffini, J.L.& Zalucki, J.M., 2000, 'Environmental Flow Requirements of the Brisbane River Downstream of Wivenhoe Dam', Griffith University: Brisbane, Australia: South East Queensland Water Corporation, and Centre for Catchment and In-Stream Research.

Arthington, A.H., Tharme, R.E., Brizga, S.O., Pusey, B.J. & Kennard, M.J., 2004, 'Environmental flow assessment with emphasis on holistic methodologies', in R. L. Welcomme, & T. Petr, eds, 'Proceedings of the Second International Symposium on the Management of Large Rivers for Fisheries, Volume 2: Sustaining Livelihoods and Biodiversity in the New Millennium', Phnom Penh, Kingdom of Cambodia,11-14 February 2003. Bangkok, Thailand: FAO, RAP, 37-65 pp. (RAP publication 2004/17)

Arthington, A.H., Naiman, R.J., Mcclain, M.E. & Nilsson, C., 2010, 'Preserving the biodiversity and ecological services of rivers: new challenges and research opportunities', *Freshwater Biology*, 55, 1-16. doi:10.1111/J.1365-2427.2009.02340.

Atkins, H. & Humphreys, H., 2013, 'Provision Of Consultancy Services For Preparation Of an Integrated Water Resources Management and Development Plan for the Ruvuma River and Southern Coast Basin', Component 1, Volume 3-Water Demand Assessment, 144 pp.

Barrow, C., 1998, 'River basin development planning and management-A critical review', *World Development*, 26 (1), 171-186.

Beilfuss, R.D., 2001, 'Hydrological disturbance, ecological dynamics, and restoration potential: the story of an African floodplain', Dissertation (PhD), University of Wisconsin-Madison.

Beilfuss, R.D., Dutton, P. & Moore, D., 2000, 'Land cover and land uses in the Zambezi Delta', Volume III, Chapter 2, Land Cover Change Zambezi Delta, in J. Timberlake, ed., Biodiversity of the Zambezi Basin wetlands, Occasional Publications in Biodiversity No. 8, Biodiversity Foundation for Africa, Bulawayo, Zimbabwe.

Beilfuss, R. & Brown, C., 2006, 'Assessing environmental flow requirements of the Zambezi Delta: Application of the DRIFT model (Downstream Response to Imposed Flow Transformations)', Maputo, Mozambique: University of Eduordo Mondlane.

Beilfuss, R. & Brown, C., 2010, 'Assessing environmental flow requirements and trade-offs for the Lower Zambezi River and Delta, Mozambique', *International Journal of River Basin Management*, 8, 127-138, DOI: 10.1080/ 15715121003714837

Bovee, K.D., 1982, A Guide to Stream Habitat Analysis Using the Instream Flow Incremental Methodology, Instream Flow Information Paper 12. FWS/OBS-2/26. Washington, D.C: USDI Fish and Wildlife Services, Office of Biology Services.

Brisbane Declaration, 2007, Environmental flows are essential for freshwater ecosystem health and human well-being', 10th International River Symposium and Environmental Flows Conference, Brisbane, Australia, 3-6 September, 2007.

Chimatiro, S.K., 2004, 'The biophysical dynamics of the Lower Shire River floodplain fisheries in Malawi', Rhodes University, Grahamstown: Unpublished PhD thesis, 259 pp.

Chenje, M., 2000, State of the Environment 2000. Zambezi Basin, Maseru/Harare/ Lusaka: SADC/IUCN/ZRA/SARDC 2000.

Cohen, L.T., 2013, 'Influence of dam operation on water resources management under different scenarios in the Zambezi River Basin considering socio-economical and environmental constraints', PhD Thesis, École Polytechnique Fédérale de Lausanne, Lausanne, Switzerland.

Costanza, R., D'arge, R., De Groot, R., Farber, S., Grasso, M., Hannon, B., Limburg, K., 1997, The value of the world's ecosystem services and natural capital. Nature, 387, 253-260.

Crowder, D.W. & Diplas, P., 2000, 'Using two-dimensional hydrodynamic models at scales of ecological importance', *Journal of Hydrology*, 230, 172-191.

Da Silva, J., 1986, River runoff and shrimp abundance in a tropical coastal ecosystem-the example of the Sofala Bank Assessing environmental flow requirements and trade-offs for the Lower Zambezi River and Delta (Central Mozambique), NATO ASI Series 67, Berlin, Springer Verlag, pp. 329-344.

Davies, B.R., Beilfuss, R.D. & Thomas, M.C., 2000, 'Cabora Bassa retrospective, 1974-1997: effects of flow regulation on the Lower Zambezi River', *Verh International Verein Limnology*, 27, 1-9.

Denconsult, 1998, Sector Studies under ZACPLAN, Sector Study No. 7.Environmental Impact of Expanding Utilization of Water Resources from the Basin, Volume I, Final Report. Lusaka, Zambia: Zambezi River Authority.

Dunbar, M.J., Gustard, A., Acreman, M.C. & Elliott, C.R.N., 1998, 'Review of Overseas Approaches to Setting River Flow Objectives', Environment Agency R&D Technical Report W6B 96, Wallingford, UK: Institute of Hydrology.

Epsmo-Biokavango E-Flows Team, 2009, 'Hydrology report: data and models', Report 05-2009, EPSMO/BIOKAVANGO, Okavango Basin Environmental Flows Assessment Project, Maun, Botswana: OKACOM.

Gammelsrød, T., 1992, Variation in shrimp abundance on the Sofala Bank, Mozambique, and its relation to the Zambezi River runoff. *Estuarine, Coastal and Shelf Science*, 35, 91-103.

GLOWS-WRBWB\EFA\WAMI, 2013, 'Wami River sub-basin, Tanzania. Initial Environmental Flow Assessment, Phase II. Wet season data collection and corresponding revisions to environmental flow recommendations from 2007', 93 pp.

GLOWS-FIU, 2012, 'Environmental Flow Recommendations for the Mara River, Kenya and Tanzania', Miami, Florida: Global Water for Sustainability program (GLOWS).

GLOWS-FIU, 2014, Environmental Flow Recommendations for the Ruvu River Basin, ISBN-13: 978-1-941993-02-6, 52 pp. http://www.tzdpg.or.tz/fileadmin/documents/dpg_internal/dpg_working_groups_clusters/cluster_2/water/WSDP/WRM_-_Component_1/Ruvu_EFA_Final_Print.pdf. [Viewed 28 July 2016]

GOM, Government of Malawi, 2015, 'Water Resources Investment Strategy. Component 1-Water Resources Assessment: Annex II-Surface Water. Ministry of Irrigation and Water Development. Second National Water Development Project (NWDP II)

Hoguane, A.M., 2000, 'The role of Zambezi runoff in the shrimp abundance in Sofala Bank', Maputo, Mozambique: Eduardo Mondlane University, 15 pp.

Isaacman, A. & Sneddon, C., 2000, 'Toward a social and environmental history of the building of Cabora Bassa Dam', *Journal of Southern African Studies*, 26, 597-632.

Jacimovic, R. & O'keeffe, J., 2008, 'A Guide for deciding on levels of detail for Environmental Flow Assessment', Project reference No 9E0752.01, Report for WWF Mediterranean, 32 pp.

Jowett, I.G., 1997, 'Instream flow methods: A comparison of approaches'. *Regulated Rivers: Research & Management*, 13, 115-127.

Kamweneshe, B., Beilfuss, R. & Simukonda, C., 2002, 'Population and distribution of Kafue Lechwe and other large mammals on the Kafue flats, Zambia. Zambezi Basin Crane and Wetland Conservation Program', Working Paper #6.

Kashaigili, J.J., Mccartney, M. & Mahoo, H.F., 2007, 'Estimation of environmental flows in the Great Ruaha River catchment, Tanzania', *Physics and Chemistry of the Earth,* Parts A/B/C, 32, 1007-1014.

Kashaigili, J.J., Mccartney, M.P., Mahoo, H.F., Lankford, B.A., Mbilinyi, B.P., Yawson, D.K. & Tumbo, S.D., 2006, Use of a hydrological model for environmental management of the Usangu Wetlands, Tanzania, Colombo, Sri Lanka: *International Water Management Institute* (IWMI).

King, J.M., Tharme, R.E. & Brown, C.A., 1999, 'Definition and Implementation of Instream Flows', Thematic Report for the World Commission on Dams. Southern Waters Ecological Research and Consulting: Cape Town, South Africa.

King, J.M. & Tharme, R.E., 1994, Assessment of the Instream Flow Incremental Methodology and Initial Development of Alternative Instream Flow Methodologies for South Africa. Water Research Commission Report No. 295/1/94. Pretoria, South Africa: Water Research Commission.

King, J.M. & Louw, M.D., 1998, 'Instream flow assessments for regulated rivers in South Africa using the Building Block Methodology', *Aquatic Ecosystem Health and Management* 1, 109-124.

King, J., Brown, C. & Sabet, H., 2003, 'A scenario-based holistic approach to environmental flow assessments for rivers', *River Research and Applications* 19: 619-639.

King, J.M., Tharme, R.E. & De Villiers, M.S., Eds, 2008, Environmental Flow Assessments for rivers: Manual for the Building Block Methodology (Updated Edition) WRC Report No TT 354/08.

King, J.M., Brown, C.A. & Barnes, J., 2009, 'Final IFA Project Report', Report 08-2009 EPSMO/BIOKAVANGO Okavango Basin Environmental Flows Assessment Project, OKACOM, Maun, Botswana, 44 pp.

Kunz, M.J., Anselmetti, F.S., West, A., Wehrli, B., Vollenweider, A., Thüring, S. & Senn, D.B., 2011, 'Sediment accumulation and carbon, nitrogen, and phosphorus deposition in the large tropical reservoir Lake Kariba (Zambia/Zimbabwe)', *Journal of Geophysical Research G: Biogeosciences*, 116, G03003.

Kleynhans, C.J., 1996, 'A qualitative procedure for the assessment of the habitat integrity status of the Luvuvhu River (Limpopo system, South Africa)', *Journal of Aquatic Ecosystem Health*, 5,41-54.

Lagerblad, L., 2011, 'Assessment of environmental flow requirements in Buzi Riverbasin, Mozambique', Uppsala, Sweden: Department of Ecology and Environmental Science, Uppsala University, 69 pp.

Macdonald, M., 2007, 'Integrated Water Resources Management Strategy for the Zambezi River Basin. Rapid Assessment-Final Report', Norwegian Embassy Lusaka, Zambia:SADC-WD/Zambezi River Authority, CIDA, DANIDA, 206 pp.

Manatunge, J., Nakayama, M. & Priyadarshana, T., 2008, 'Environmental and social impacts of reservoir: issues and mitigation', *Oceans and aquatic ecosystems*, 1,212-255.

Mazvimavi, D., Madambombe, E. & Makurira, H., 2007, 'Assessment of environmental flow requirements for river basin planning in Zimbabwe', *Physics and Chemistry of the Earth*, Parts A/B/C, 32, 995-1006.

Mccartney, M., Cai, X. & Smakhtin, V., 2013, 'Evaluating the flow regulating functions of natural ecosystems in the Zambezi River Basin', Colombo, Sri Lanka: International

Water Management Institute (IWMI). 59pp. (IWMI Research Report 148). doi:10.5337/2013.206

Milhous, R.T., Updike, M.A. & Schneider, D.M., 1989, 'Physical Habitat Simulation System Reference Manual-Version 2. Instream Flow Information Paper 26. USDI Fish and Wildlife Services, Washington DC, US', Biology Report, 89(16), 1-403.

MoW, 2015, 'Environmental Water Requirements Assessment Guidelines for Tanzania-Draft 1', Dar es Salaam, Tanzania: Ministry of Water and Irrigation, 106pp.

Moyo, N.A.G., 1994, 'An analysis of fish yield predictions in African lakes and reservoirs', *African Journal of Ecology*, 32, 342-347.

Naiman, R.J., Bunn, S.E., Nilsson, C., Petts, G.E., Pinay, G. & Thompson, L.C., 2002, 'Legitimizing fluvial ecosystems as users of water', *Environmental Management*, 30, 455-467.

Nindi, S.J., 2007, 'Changing livelihoods and the environment along Lake Nyasa, Tanzania', *African Study Monographs*, 36, 71-93.

PBWO/IUCN, 2009, 'Pangani River Basin Flow Assessment: Final Project Report', Pangani Basin Water Board, Moshi and IUCN Eastern & Southern Africa Regional Program. 89 pp.

Poff, N.L & Ward, J.V., 1989, 'Implications of streamflow variability and predictability for lotic community structure: a regional analysis of streamflow patterns', *Canadian Journal of Fisheries and Aquatic Sciences*, 46, 1805-1818.

Poff, N.L., Allan, J.D., Bain, M.B., Karr, J.R., Prestegaard, K.L., Richter, B.D., Sparks, R.E. & Stromberg, J.C., 1997, 'The natural flow regime: A paradigm for river conservation and restoration', *BioScience* 47: 769-784.

Poff, N.L., Richter, B.D., Arthington, A.H., Bunn, S.E. & Naiman, R.J., 2010, 'The ecological limits of hydrologic alteration (ELOHA): A new framework for developing regional environmental flow standards', Freshwater Biology, 55, 147-170. Doi:10.1111/J.1365-2427.2009. 02204.X

Postel, S. & Carpenter, S., 1997, 'Freshwater ecosystem services', in Daily, G.C., ed., Nature's services; societal dependence on natural ecosystems, Washington D.C, USA: Island Press.

Prewitt, C.G. & Carlson, C.A., 1980, Evaluation Of Four Instream Flow Methodologies Used On The Yampa And White Rivers, Colorado, Denver, Colorado: Biological Sciences Series Number 2, Bureau Of Land Management.

Pusey, B.J., 1998, 'Methods Addressing the Flow Requirements of Fish', in A. H. Arthington, & J. M. Zalucki, Eds, Comparative Evaluation of Environmental Flow Assessment Techniques: Review of Methods, Occasional Paper No. 27/98. Canberra, Australia: Land and Water Resources Research and Development Corporation, pp. 66-105.

Rapport, D.J., Costanza, R. & Mcmichael, A.J., 1998, 'Assessing Ecosystem Health', *Trends in Ecology And Evolution*, 13, 397-402.

Richter, B.D., Baumgartner, J.V., Powell, J. & Braun, D.P., 1996, 'A Method For Assessing Hydrological Alteration Within Ecosystems', *Conservation Biology*, 10(4), 1163-1174.

Richter, B.D., Baumgartner, J.V., Wigington, R. & Braun, D.P., 1997, 'How Much Water Does A River Need?' *Freshwater Biology*, 37, 231-249.

Rosenberg, D.M., Mccully, P. & Pringle, C.M., 2000, 'Global-Scale Environmental Effects of Hydrological Alterations: Introduction', Bioscience, 50 (9), 746-751. SADC, (Southern African Development Community), 2000, Revised Protocol on Shared Watercourse Systems. Http://Www.Sadc.Int/Files/3413/6698/6218/ Revised_Protocol_On_Shared_Watercourses_-_2000_-_English.Pdf. [Accessed On 29 January 2016].

Scholes, R.J. & Biggs, R., 2004, Ecosystem Services In Southern Africa: A Regional Assessment, Pretoria, South Africa: Council For Scientific And Industrial Research.

Scott, W.P., 2003, Integrated Kafue River Basin Environmental Impact Assessment Study. State of the Environment Report, Lusaka, Zambia: Ministry of Energy and Water Development, Opppi/Zesco Limited.

Smakhtin, V.Y., Revenga, C. & Döll, P., 2004, Taking Into Account Environmental Water Requirements In Global-Scale Water Resources Assessments. Comprehensive Assessment Research Report 2. Colombo, Sri Lanka: Comprehensive Assessment Secretariat.

SMEC, 2013, Preparation of an Integrated Water Resources Management and Development Plan for Lake Nyasa Basin, Final Interim 2 Water Resources Assessment Report, 131 Pp.

Stalnaker, C.B., 1982, 'Instream Flow Assessments Come of age in the Decade of the 1970s', In: W. T. Mason, & S. Iker, Eds, Research on Fish and Wildlife Habitat. Epa-600/8-82-022. Office of Research and Development. Washington, Dc: Us Environmental Protection Agency, 119-142.

Stalnaker, C.B., Lamb, B.L., Henriksen, J., Bovee, K.D. & Bartholow, J. 1994, The Instream Flow Incremental Methodology: A Primer for Ifim. National Ecology Research Centerinternal Publication, Fort Collins, Co, Usa: National Biological Survey.

Sushka, J. & Napica, P., 1986, 'Ten Years After Completion of Cahora Bassa Dam. In: Unep/ Unesco, Eds, The Impact of Large Water Projects on the Environment: Proceedings of an International Symposium, Paris: Unep/ Unesco, 171-203.

Tharme, R.E., 1996, Review of International Methodologies for the Quantification of the Instream Flow Requirements of Rivers, Water Law Review Final Report for Policy Development for the Department of Water Affairs and Forestry, Pretoria. University of Cape Town, South Africa: Freshwater Research Unit.

Tharme, R.E. & King, J.M., 1998, 'Development of the Building Block Methodology for Instream Flow Assessments, and Supporting Research on the Effects of Different Magnitude Flows on Riverine Ecosystems', Water Research Commission Report No. 576/1/98, 452 Pp.

Tharme, R.E., 2000, 'An Overview Of Environmental Flow Methodologies, With Particular Reference To South Africa', in J.M. King , R. E. Tharme & M. S. De Villiers eds, Environmental Flow Assessments for Rivers: Manual for the Building Block Methodology, Water Research Commission Technology Transfer Report No. Tt131/00, Pretoria, South Africa: Water Research Commission, 15-40.

Tharme, R.E., 2003, 'A Global Perspective on Environmental Flow Assessment: Emerging Trends in the Development and Application of Environmental Flow Methodologies for Rivers', River Research Applications, 19, 397-441.

Trihey, E.W. & Stalnaker, C, B., 1985, 'Evolution and Application of Instream Flow Methodologies to Small Hydropower Developments: An Overview of the Issues', In: F. W. Olson, R. G. White, & R. H. Hamre, eds, Proceedings of the Symposium on Small Hydropower and Fisheries, Aurora, Colorado.

Turner, B., 1984, 'The Effect of Dam Construction on Flooding of the Kafue Flats', pp. 1-9, In: W.L. Handlos, & G. Williams, eds, 'Development of the Kafue Flats-The Last Five Years', Lusaka: Kafue Basin Research Committee of the University Of Zambia.

Turpie, J., Smith, B., Emerton, L. & Barnes, J., 1999, 'Economic Value of the Zambezi Basin Wetlands'. Cape Town, South Africa: IUCN Regional Office For Southern Africa, 346 pp.

Tweddle, D., Makwinja, R.D. & Sodzapanja, G., 1995, 'Catch and Effort Data for the Fisheries of the Lower Shire River and Associated Marshes, 1976-1993', *Malawi Fish. Bulletin*, 31, 126 pp.

URT-United Republic of Tanzania, 2002, National Water Policy, Ministry of Water and Livestock Development, Dar es Salaam.

URT-United Republic of Tanzania, 2009, Water Resources Management Act, Dar Es Salaam, Tanzania: Ministry of Water.

URT-The United Republic of Tanzania, 2010, 'The Tanzania Development Vision 2025', Dar es Salaam, Tanzania: Planning Commission.

Waddle, T., 1998, 'Development of 2-Dimensional Habitat Models', in S. Blaz'Kova', C. Stalnaker, & O. Novicky', eds, Hydroecological Modelling. Research, Practice, Legislation and Decision making, Report by US Geological Survey, Biological Research Division and Water Research Institute, Czech Republic, Vuv, Praha: Fort Collins, and Water Research Institute, 19-22.

Welcomme, R.L., 1985, 'River Fisheries', FAO Fisheries Technical Paper, 262, 330 Pp. Wrem, 2014, 'Lake Rukwa Basin Integrated Water Resources Management and Development Plan. Final Report', Environmental Flow Assessment for Lake Rukwa Basin. 136 Pp.

WWF Tanzania Country Office (WWF-TCO), 2010, 'Assessing Environmental Flows for the Great Ruaha River, and Usangu Wetland', Tanzania. WWF Tanzania Country Office, 274 pp.

ZAMCOM, SADC, SARDC, 2015, 'Zambezi Environment Outlook 2015. Harare, Zimbabwe, Gaborone, Botswana', ZAMCOM, SADC, SARDC.

ZEMA & UNEP, 2012, Zambia Atlas of Our Changing Environment, Grid-Sioux Falls, Lusaka: ZEMA, UNEP Grid-Arendal.

ZINWA-Zimbabwe National Water Authority, 1996, 'Chapter 20:25-Zimbabwe National Water Authority', Harare, Zimbabwe: Zimbabwe Government Printers.

Zurbrügg, R., Wamulume, J., Kamanga, R., Wehrli, B. & Senn, D.B., 2012, 'River-floodplain exchange and its effects on the fluvial oxygen regime in a large tropical river system (Kafue Flats, Zambia)', *Journal of Geophysical Research*, 117, G03008.

8

A Synthesis of the Ecological Changes in the Zambezi River Basin: Lessons Learnt

Introduction

This book, *Ecological Changes in the Zambezi River Basin*, is a result of a collated, collaborated effort among various disciplines, research themes, think tanks (research institutions) and paradigms. The objectives of the study, as expounded in Chapter 2 of the book revolve around the comparability of sub-catchments within the ZRB, focusing on ecological changes, biological monitoring, climate, hydrology and environmental flows as the major research themes. Interwoven into these themes are cross-cutting themes to do with livelihoods, indigenous knowledge and policy which are given more scrutiny at sub-catchment level and, where possible, trends and comparisons are made across the basin. This, in essence, was our conceptual framework and the analytical approach used throughout the study is the comparative research method. The highlights, insights and salient points from the study and across the book chapters are presented herein as lessons learnt and a platform from which future research can build upon.

Comparative Research

The Comparative Research Method (CRM) is an interesting approach to scientific inquiry that can be used to decipher similarities (and differences) spatially and temporally within a given study area. One definition describes it as 'an approach that aims to compare two or more entities with the intention of discovering a specified or approximate amount expressed in relation to a specific number, quantity or quality about one or more of the units being compared' (Azarian 2011). This definition hints on the need of the CRM to incorporate precise, quantifiable similarities or differences; the use of either or both qualitative and quantitative approaches where applicable and specified comparable entities. As most researchers can relate, it is often unlikely for such a demand for quantifiable comparison to be made due to limitations of the nature and extent of comparison that can possibly be made, more so, the availability of comparable data sets. Our experience in using the CRM was quite fascinating and thought-provoking, given the large expansive and heterogeneous nature of the ZRB and, therefore, seemingly difficult comparability across the basin. These difficulties arise in; differences in topography and climate; hydrology, ecosystems and communities and the multi-disciplinary nature of our research team. The major question posed was how to compare the ZRB, given these obvious differences as well as the varied research themes (ecosystems, climate, communities, and hydrology) that we wished to pursue.

In the end, we chose to use a mixed methods approach and to make use of case studies given the multi-disciplinarity, complexity of the landscape and quite ambitious objectives we had chosen to tackle (see Chapter 2). We concentrated on the Manyame River Catchment as a result of its proximity to the project coordination as well as the availability of both primary and secondary data sources for this sub-catchment. Comparisons were then made with the other four sub-catchments using secondary data sources. Where and when data was available or availed, depending on the source, some direct comparisons across specified areas could be made such as in the case of the climate and hydrological data used in chapters 5 and 6 respectively for the different sub-catchments within the Zambezi River Basin. More often, however, data of the same nature that covered similar or comparable areas and collected in the same manner in the ZRB were unavailable. In such cases, reviews and analyses of what has been covered in terms of the development in research, findings, gaps and suggested way forward, such as in the case of the use of biological monitoring and environmental flows (chapters 4 and 7), were then presented.

Research gaps, missing data and dissimilarities in methodologies are part of the quandary of the CRM. Therefore, the use of the CRM becomes both an art in the sense of comparing, and a science in the conceptualization and methodological approach. Both qualitative and quantitative methods of inquiry were used in this study in order to ensure community and stakeholder participation and to elicit perceptions on weather, climate, natural resources and livelihoods. For the qualitative assessments, we managed to interrogate two communities in the Manyame River Catchment to elucidate perceptions and the use of local indigenous knowledge systems and practices (LIKSP). This, of course, not only fell in tandem with our objectives, but resonated well with our approach and insistence not to sideline communities as is often done by disciplinary studies, natural sciences in particular. The results of such an integration of disciplines came out clearly in Chapters 3 and 5 where the two respective case studies of community perceptions are presented. This integration of quantitative and qualitative data was important as our definition of ecological changes encompassed not only the physical ecosystems but also the social aspects and, more specifically, communities living within the ecosystems of the ZRB. These are the very people who interact, use and largely depend on the ZRB; and therefore it was important for us to capture their thoughts and perspectives.

In this regard, our approach to the CRM was not fixed but made use of both quantitative data (climate, hydrology, biological monitoring and ecological changes) and qualitative data in a complementary manner. Therein lies the mixed methods approach, the fusion of different disciplines (inter and/or multi-disciplinarity) as well as complementaries. It was not an easy path to follow and was often accompanied by heated arguments, misunderstandings and provocation for us to consistently inquire more, probe further and venture into unfamiliar territories as interfaces of the different disciplines represented. We conclude that the CRM is a useful cognitive tool, albeit there is need for careful, contemplative reflection, in the approaches in which the CRM can be used and if at all extrapolated.

Lessons Learnt

On Commonality of Communities and Resources for the Five Sub-Catchments

The five sub-catchments within the ZRB under study were varied in nature. The communities and natural resources therein also d-iffered. Despite this, and as expounded in Chapter 2, it is apparent that all the five sub-catchments are experiencing increasing environmental pressure on their resources due to increasing human populations and, therefore, high dependence on the river ecosystems, which in some cases was quoted as 'extreme'. This was the case of the Shire Catchment, whose land degradation in some parts of the basin was said to be severe mainly as a result of poor resource management and high population densities. Similarly, both the Manyame and Songwe catchments were shown to lie in densely populated districts, while significant, continued increases in the population of the ZRB are expected. Thus, the human population factor stands out as glaring vis-à-vis the natural resources base, and especially water resources in the ZRB. The growing population and subsequent increasing demand of resources on the ZRB are highlighted in chapters 1, 3 and 6, and also as part of the drivers-pressures-states-impacts-responses (DPSIR) model that we used (see Chapter 3).

An interesting observation was the decreased dependency of communities on land resources versus river resources highlighted in the Manyame case study. This points to the ever-shifting natural resource base and its returns to communities, among a myriad of factors. One of the strong factors, as pointed out in this volume, is land use patterns (see chapters 2 and 4) which, despite varying landscapes among the five sub-catchments, were common across them. Anthropogenic activities cited within the case studies included agriculture (crop land, pastures,) deforestation, damming and industrialization, development of hydropower, fisheries and forests, soil erosion and sedimentation, mining and sand extraction, among other factors. Increase in these anthropogenic activities and production processes creates a propensity towards limitation of and strain on the ecosystem goods and services provided by the ZRB. In turn, this forms potential threats for increase in undesirable levels of elements such as sediments, nutrients and pollutants, which can result in a decrease in water quality as well as ecosystem services and functioning. Therefore, land use patterns, management strategies and extractive processes need to be addressed if the proper management of the ZRB is to be sustainable and if the livelihoods of the ZRB communities, especially fisheries, are to be protected.

Ecological Changes

Several ecological changes of the ZRB are cited in the study and chief among these were the undercutting of the Zambezi River channel, dwindling water resources as well as reductions in the fish resource base, among other changes. The drivers of these ecological changes are mostly a result of anthropogenic causes and, namely demographic dynamics (human population increases) and increased demand on the ZRB for agricultural land and water resources, human-induced climate changes, mining, urbanization and the introduction of new (exotic) species. Of interest to note are the decreases in the natural fish resources in the ZRB and the East and southern African region in particular and yet communities therein still rely heavily on these resources as explained before. The nexus between ecosystems, livelihoods and agriculture, interwoven with climate as one of the major environmental factors (as a provider of water for agriculture and other livelihood uses) is noted in the study. Therefore, ecological changes have huge implications for livelihoods, given the ZRB communities' dependence on the system for hydropower, water resources and fisheries, as some of the major ecosystem goods and services derived from there.

Biological Monitoring

Given the importance of the ZRB, the monitoring of the ZRB ecosystems becomes pertinent given the background of the pressures on its natural resources, as well as the precious goods and services that the basin provides. One of the ways in which effective monitoring of the physical, chemical and biological components of freshwater ecosystems can be done is through biological monitoring. Chapter Four highlights how several living organisms and namely macroinvertebrates, fish and periphyton can be used to assess the water quality and environment in and around flowing waters. In addition, we showcase how numerous biological monitoring tools have been formulated, tried and tested for the east and southern African region's flowing waters (see Chapter 4). The highlight of this review is that, for a long time, the region has depended on biological monitoring tools that have been developed for other regions, mainly the temperate region and specifically South African streams and rivers. However, the Tanzanian case study shows that, for the past decade, much progress has been made in developing locally-adapted, biomonitoring protocols for Tanzanian rivers. The development of such biomonitoring tools is advocated for Zimbabwean lentic systems where a case study of the Manyame River showed that among the three biota tested, diatom assemblages were the best indicators of environmental degradation, land use patterns and the water quality of the river, followed by macroinvertebrates, with fish being the least sensitive. Therefore, this could be a good starting point in developing bioassays for the other six major river catchments of Zimbabwe. Similarly, the progress made in the other ZRB riparian countries and the east and southern African region as a whole can be used to upscale efforts towards more refined and locally contextualized biological monitoring tools. In fact, locally-engineered biological monitoring systems must be developed for each major sub-catchment, which may mean several for the ZRB.

Climate and Livelihoods

Climate is one of the deep-seated systemic challenges to contend with across the five sub-basins, especially in terms of the vulnerability of the sub-catchments. For instance, the high dependence of the Barotse Flood Plain on the flooding pulses of the Zambezi makes the community's farming system quite vulnerable to climate change and variability (reduced water flows, increased temperature and evaporation; increasing extreme weather events such as droughts, severe storms, higher annual floods and flood failure), given their increasing nature in the area.

Historical climate trends over a 34-year period covering 1980-2014 showed a changing climate in all the five sub-catchments. The southern basins (Barotse and Manyame) have become warmer and drier over the studied 34-year period while the central and easterly sub-catchments (Shire, Songwe and Ruhuhu) were shown to have become wetter and warmer—though this warming is to a lesser extent as compared to the other two sub-catchments over the same period. Projected future climate scenarios of the same sub catchments reflected more warming and the same trends in rainfall (drier and wetter for the southern and easterly sub-catchments respectively) in the mid-future periods. This, of course, bears potential implications for future planning for the ZRB, especially in terms of water resource management.

While some of the debates around climate modelling and climate data availability are explicated in Chapter 5, the exciting part for us was the integration of the physical data analyses with local indigenous knowledge systems and practices (LIKSP), as well as the presentation of a case study on community perceptions regarding weather and climate trends in the Manyame River sub-catchment and, consequently, the community's adaptation to these phenomena. The use of the survey offered more insight not only into the adaptation aspects, but in matching perceptions to actual climate data, as well as investigating how the community perceived weather, climate and ecological changes in the context of their capita, i.e. economic, human, natural and social resources. Our study reflected that the challenges compounded by climate changes and variability are related mostly to the access and control of human, natural and physical capital assets in the ZRB. An additional eye-opener here was the fact that the ZRB and the Southern African Development Community (SADC) region in particular have many common LIKSP that can be harnessed and incorporated into more robust climate monitoring, analytical and communication systems. Notably, many research gaps exist in terms of the availability and access to climate data which limited the present study to 34 years of climate data analysis. Such study gaps must be pursued and refined if disaster risk assessments and reduction, as well as monitoring related to climate, are to be improved for the ZRB.

Hydrology

The monitoring of the hydrology of the ZRB is equally pertinent as water resources are critical for the development of the region. Not only are the waters of the ZRB important for their use in agriculture, fisheries, industries and domestic transport and navigation but, more compositely, for the natural ecosystems in and around the waters themselves. Therefore, focusing on the hydrology of the ZRB gave us a good picture of two important facets of the hydrological cycle: rainfall and water flows along parts of the basin. Analyses of the climate in Chapter 5 showed that rainfall patterns within the ZRB vary as explained earlier. The analysis of water flows at six sub-catchments (Chapter 6) which showed decreasing (albeit statistically insignificant) trends in the long term is disconcerting, especially given the projected increase in water use from the current 20 per cent to 37.32 per cent and eventually 63 to 77 per cent of the current mean annual runoff in the near and far future respectively, more particularly for the drier parts of the ZRB. This unbalanced demand for water as compared to that available from runoff is quite unsustainable for the basin and bears potential harmful impacts on the ecosystem structure and functioning with consequent, possible negative impacts for ecosystems as well as livelihoods and the millions of lives reliant on the ZRB in the future. At the same time, the imbalance, though posing constraints, also presents opportunities to harness the water resources in the region and this is why we advocate for the integrated water resources management approach as a holistic tool and best way to attempt to realise sustainable utilization of the water resources in the ZRB.

Environmental Flows

The hydrological analyses could not be complete without us considering environmental flows (EFs) within the ZRB. This is because EFs are one of the most prominent metrics used to measure the health of flowing water systems and present a threshold that must not be crossed if the aquatic systems are to be maintained as sustainable and able to continue providing the all-important ecosystem goods and services. In the present study, a review of EFs within the ZRB showed that there are a number of approaches used in monitoring EFs within the ZRB. Therefore, it is not easy to compare EFs across the basin in view of the different approaches and the uncertainties of the estimates currently used. There is therefore need to refine methods of EF assessment as data and more information becomes available in the sub-basins. However, an average use of 10-30 per cent of mean annual runoff (MAR) is recommended for rivers across the basin for the dry season and 30-60 per cent during the wet seasons. We conclude this section of the book by discussing the fact that for real investments in environmental flow assessments to be made, there needs to be a justification and demonstration of improvements in environmental, social and economic conditions (sustainability model) within the wider society, as opposed to the few, scattered ones demonstrated by isolated actors or donor efforts.

Policy, Legal and Institutional Issues

Many topical issues and concerns have been revealed in our study and the lessons learnt highlighted in this chapter. In addition, several authors (Chenje 2000, Hirji et al 2000, Zambezi Environmental Outlook 2015, Tumbare 2008) discuss the importance of policy, legal and institutional development of the ZRB. If any of our recommended steps and actions as well as those from related works and others are to work, there is need for great, concerted efforts of policy, regulation (governance), legal and institutional development to safeguard the ZRB. Not that there are neither policies and regulatory frameworks nor any legal and institutional structures currently in place, but our study demonstrates how these often tend to be fragmented and uncoordinated over a transboundary resource (see chapters 6 and 7). All the eight riparian countries of the ZRB have policies and laws, governance and institutional agencies, which recognize the importance of the Zambezi River and its resources, sub-catchments and the need to protect them. However, the major let-down is in the actual implementation or incorporation in these policies or, even worse, the splintered way in which the policies are packaged and put into practice, as well as the way they do not necessarily fit into the regional frameworks. As an example, even though much effort is being made to integrate climate (change) policies and strategies at country level through the United Nations Framework Convention on Climate Change (UNFCCC)-led initiative, the reality is that such phenomena as climate know no arbitrary, man-made country boundaries across the very, and with the very, transboundary ZRB. Therefore, an eco-regional approach to protecting and policing resources such as the ZRB must be adopted not only at regional level but one that filters down to country, sub-catchments and other local levels. We advocate for a regional approach that is not top-down but rather built from bottom up and synchronized to reach the regional level of cooperation (see Chapter 7).

There are several regional efforts that have been made in institutional development and management of the ZRB but we still have a long way to go. Operational regional bodies include water-specific institutions, management bodies and programmes such as the Joint Permanent Water Commission (JPWC) between Botswana and Namibia, SADC Regional Water Policy, SADC Protocol on Shared Waters and Systems, Zambezi System Action Plan (ZACPLAN), the Zambezi River Authority (ZRA) between Zambia and Zimbabwe and the Zambezi Commission (ZAMCOM); which were commissioned by SADC and, to a smaller extent, the African Union (AU) structures. In line with the protocol of shared water courses, the Zambezi River is managed by ZAMCOM, which is specifically responsible for the management of the ZRB. In addition, other, related, (global) conservation programmes such as CITES, RAMSAR, Convention on Biological Diversity (CBD) and Convention to Combating Desertification (CCD), to mention a few, are also operational within the policy, legal and institutional framework of the ZRB. However, the mandates of regional bodies are often misconstrued and disjointed, while those of the international level are often weak in implementation at the local level. ZRA, for instance, although operating under the SADC Shared Water Protocol which is quite encompassing, is concerned more with water abstraction and withdrawal between the Zambia-Zimbabwe shared portion of the Zambezi than other facets of water management such as water conservation or the maintaining of environmental water flows for the protection of ecosystem functioning, which is a fascination of an isolated programme led by the World Wide Fund for Nature (WWF) in some parts of the ZRB.

This then leaves gaps in the current management approach(es) in practice across the ZRB. Hirji et al. (2000) put it this way: '…the concept of sustainable water resources management is poorly understood by policy makers, water resources planners and managers'. Sixteen years down the line, it is sad to note that this misconception still exists and reflects a gap in human capacity and perhaps institutional development as well as coordinated efforts at regional level. As an example, a review of the policies and legal frameworks across the ZRB showed differences as well as an inaccurate application of the SADC Regional Water Policy, with the exception of Tanzania. This regional policy adopts integrated water resources management (IWRM) principles and approaches but this is not the case with the country-specific approaches. This is because individual countries have isolated initiatives and plans for the use, management and development of water resources of the Zambezi River in the in-country portions of the river with little, if at all any, regard for downstream effects and the regional consensus and efforts which have been developed over the past 30 years.

Conclusion

In conclusion, this study has shown that the ZRB is experiencing a number of ecological changes mainly caused by anthropogenic activities and the consequent increasing demand on the basin from an ever-increasing human population. In addition, the climate of the five understudied sub-catchments within the ZRB is changing and, together with increasing climate variability, consequently compounds the ecological changes. This leads to adverse impacts on water resources, fisheries and the ZRB and its tributaries' health and integrity as well as the livelihoods of the communities therein. Communities are coping and adapting to the ecological changes through the use of local indigenous knowledge and in some cases by shifting livelihood dependence on land resources to water-based resources as reported in some parts of the basin. We also highlight that there are many monitoring programmes and tools in place for determining the health (biological integrity, water quality and quantity) of the ZRB which include biological monitoring protocols as well as water flow measurements and environmental flow assessments. However, there is need for the improvement of these efforts as well as coordination among the different stakeholders if the basin is to be managed sustainably. One example is the integration of climate sciences with indigenous knowledge systems in effectively working with communities on the ground in the ZRB and beyond. We advocate for the use of the integrated water resources management approach as one effective and holistic tool that can be adopted at local, national and regional levels for the sustainable management of the ZRB in view of the underscored ecological changes.

When it comes to policy formulation and implementation, we conclude that there is a need for a holistic approach which we cannot overemphasize. This emphasizes the importance of the whole (ZRB) and the organic and functional relationship-interdependence of its parts which in the case of this study were ecosystems, climate, hydrology and communities. Of course, such focus requires much more concerted efforts than what is currently shown, and greater commitment towards the management and sustainable utilization of the ZRB from local to regional levels. For this to happen, many isolated actors, stakeholders and the ZRB riparian governments need to come together and provide solutions which reflect social and economic benefits to validate financial investments and financing for the required changes in policy (re)formulation and implementation. Showcasing the economic value of natural resources such as the

ZRB is something that we, as researchers often fail to demonstrate; but our hope is that this book, in the very least reflects the invaluable worth of the basin. Underlying all this are opportunities for innovation, technologies and sustainable solutions to the Zambezi's future, needed sooner rather than later. And that, perhaps, will be the topic of our next book!

References

Azarian, R., 2011, 'Potentials and Limitations of Comparative method in Social Science', *International Journal of Humanities and Social Science*, 1, 113-125.

Chenje, M., 2000, State of the environment Zambezi Basin 2000, Harare, Zimbabwe: IUCN, SADC.

Hirji, R., Johnson, P., Maro, P. & Matiza_Chiuta, T., eds, Defining and mainstreaming environmental sustainability in water resources management in southern Africa, SADC, IUCN, SARDC, World Bank: Maseru/Harare/Washington DC.

Tumbare, M. J., 2008, 'Managing Lake Kariba sustainably: threats and challenges', Management of Environmental Quality: An International Journal, 19, 731-739. ZAMCOM, SADC & SARDC, 2015, Zambezi Environment Outlook 2015, Harare, Gaborone.

Index